# METHODS FOR STUDYING
# THE ECOLOGY OF
# SOIL MICRO-ORGANISMS

IBP HANDBOOK No. 19

# Methods for Studying the Ecology of Soil Micro-organisms

Edited by
## D. PARKINSON
## T. R. G. GRAY
## S. T. WILLIAMS

INTERNATIONAL BIOLOGICAL PROGRAMME
7 MARYLEBONE ROAD, LONDON NW1

BLACKWELL SCIENTIFIC PUBLICATIONS
OXFORD AND EDINBURGH

© 1971 by International Biological Programme
Published by Blackwell Scientific Publications
5 Alfred Street, Oxford, England and
9 Forrest Road, Edinburgh, Scotland

ISBN 0 632 08260 7

First Published 1971

Distributed in the U.S.A. by
F. A. Davis Company, 1915 Arch Street,
Philadelphia, Pennsylvania

Printed in Great Britain by
Adlard & Son Ltd, Bartholomew Press, Dorking
and bound at the
Kemp Hall Bindery, Oxford

# Contents

# Foreword

The International Biological Programme is a world study of 'biological productivity and human adaptability', started in 1964 and lasting for a decade until 1974. Being a wide ranging programme it is divided into seven sections, of which the one concerned with terrestrial productivity (PT) has sponsored this present handbook.

For section PT there are already five other handbooks which provide guidance and advice on methods of research. They are on the primary production of forests (No. 2), on grasslands (No. 6), on the productivity of large herbivores (No. 7), on other terrestrial animals (No. 13), and on quantitative soil ecology (No. 18). This one on the ecology of soil micro-organisms is concerned with one of the most important branches of ecology, responsible for energy flow through the ecosystems of all biome or habitat types. It should be mentioned that two other IBP handbooks which have been sponsored by other sections also deal with the ecology of micro-organisms. One of these, from section PP, deals with root-nodule bacteria (No. 15), and the other which is still in the press is concerned with the bacteria of freshwater ecosystems.

The microbiology of natural ecosystems is a subject which is advancing rapidly, and is fraught with peculiar difficulties in its methodology. Even the problem of estimating biomass, which is relatively easy in most groups of organisms, is highly complex when it comes to bacteria or minute hyphae or fungi ranging through the soil. Thus the appearance of this book, which should help materially towards the comparability of results derived from different ecosystems in different parts of the world, is timely. It will be used not only during the remaining years of IBP, but doubtless for a considerable period thereafter. Indeed, as with all the other IBP handbooks, it is hoped that experience in its use will lead to the improvement of methods and so to a new, revised edition of the book before many years have passed.

Of the three authors, Professor D.Parkinson is Head of the Department of Biology and Professor of Microbiology in the University of Calgary, Alberta,

Canada. He is also the Theme Co-ordinator for decomposition processes in Section PT of the IBP, and is currently engaged in research on soil fungi as part of the large Canadian grassland IBP project at Matador, and also on soil micro-organisms and decomposition processes as part of the Canadian tundra IBP project, Devon Island. Dr T.R.G.Gray who was a student at Nottingham University is lecturer in Botany at the University of Liverpool (1960–1971) during which period he held a post-doctoral fellowship in soil science at the University of Minnesota for a year. He was editor (with Professor Parkinson) of the *Ecology of Soil Bacteria* (Liverpool University Press 1968) and author (with Dr Williams) of *Soil Micro-organisms* (Oliver & Boyd 1971). Dr S.T.Williams is lecturer in Botany at the University of Liverpool where he was formerly a student; he conducts research in soil microbiology, especially on actinomycetes. We are much indebted to these three biologists for a valued addition to the IBP handbook series.

August 1971                                                    E.B.Worthington

# Acknowledgments

We wish to thank all those who helped us in the preparation of this manual which was made possible by the award of a NATO research grant (No. 334) for collaborative work in soil microbiology. We received many helpful ideas from our colleagues in the International Biological Programme both in the United Kingdom and Canada.

Of the many individuals who helped us, we would particularly like to thank Mrs Linda Whelan, who collected much of the information used in the book and helped to prepare the initial draft manuscript, and Mr John Bissett who made some constructive comments on chapter 3. Mrs V.Preuter was responsible for producing the typescript at all stages of its preparation and additional valuable help was received from Mrs W.Packer and Mrs. S.Collins.

Finally, we should like to thank all those microbiologists whose work we have quoted. Any errors in the description of their work are the responsibility of the authors of this manual.

# 1

# Objectives

In *IBP News 9*, 1967, under outlines of the PT program, a restricted number of methods were recommended as applicable to soil microbiological studies in IBP PT projects. These techniques, considered by many to be minimal for IBP studies have been discussed in IBP Handbook 18. This present handbook attempts to survey methods which are in use in soil microbiological laboratories and therefore discusses a much wider range of techniques. However, it will be evident from comments in this handbook that soil microbiologists are not yet in a position to answer some of the important questions being asked by other members of IBP PT projects, e.g., questions regarding the rate of microbial cell production in soil and on the relative rates of metabolic activity of different components of the soil microflora in soil microhabitats. Thus, it is not possible to measure cell production without destroying the natural environment, and the metabolism of the various components of the soil biota (including roots) is not sufficiently different to allow distinction between the components, except in certain special cases (e.g., nitrogen fixing organisms). Moreover, there seems little immediate prospect of solving these problems. In view of this, workers may well ask whether effort should not be restricted to more generalized assessments of biological activity in relation to the decomposer cycle, e.g., measurement of rates of input and output of substrates and metabolites. Perhaps problems of the partition of energy between the soil components should be shelved, at least in the context of IBP, until more much-needed research has been carried out.

Therefore, this handbook does not aim to provide a complete survey of methods in soil microbiology or production ecology but concentrates on methods which microbial ecologists have found useful. Complete standardization of methodology in studies of soil microbial ecology is not desirable and methods must be chosen and modified which attempt to provide data relevant to the soils and organisms being studied. It is hoped that this handbook will aid in the preliminary selection of suitable methods for ecological investigations, particularly in institutions where soil microbiological studies are not currently in progress. Workers in such institutions may have limited access to current literature and background information and so considerable detail is provided for many of the methods. In describing them, attempts have been made to provide a detailed description of the more important methods plus references to relevant literature, to outline clearly the purposes (general and specific) of each method, to provide a clear statement (wherever possible) on the uses and limitations of the methods and to discuss how the techniques may be tested.

Methods concerned specifically with root-nodule bacteria have been omitted since these are described comprehensively in IBP Handbook 15 (Vincent, 1970). Similarly, techniques concerned with the estimation of nitrogen fixation by micro-organisms have not been described here since they are being dealt with by workers in the Production Processes Section of IBP.

# 2

# Habitat Description

## 2.1 Selection of site

The planning of a project and its sampling program must be preceded by a proper survey of the site under study, and the collection of data on the environment and its changes. Some of the considerations necessary in the selection of forest sites are given in IBP Handbook No. 2 (Newbould, 1967).

## 2.2 Selection of sampling areas

In considering a range of terrestrial ecosystems (forest, grassland, desert, tundra) the following generalizations may be useful. Within the selected study area the need for three basic site units must frequently be recognized:

1. An undisturbed zone: the initial phases of environmental research programs are frequently exploratory, the unrestricted activities of researchers could result in the destruction of the environment under study.
2. A zone, adequate in area for multi-disciplinary research where sampling procedures are restricted to those of a non-destructive nature.
3. An extensive zone in which sampling techniques are not restricted.

Suitable entry-exit routes to the plots must be established to avoid indiscriminate movement of workers over the study area.

Following the mapping and delineation of study plots all determinations (physical, chemical and biological) should be made on a proper statistical basis.

## 2.3 Selection of environmental factors to be measured

In all studies in soil microbial ecology, investigations must normally be made on the soil(s) under study and on the environmental factors to which the microbe(s) are subjected. The amount of such data will vary according to the nature and scope of the microbiological project and the number of workers (and their analytical expertise) available to assist in its compilation.

It is impossible to give a standard list of habitat data to be collected but it must be emphasized that sufficient data should be provided to allow other workers to recognize the major features of the environment under investigation. The uncoordinated compilation of habitat data which has no relevance to the aims and applicability of the project merely represents a profligate use of available laboratory facilities.

In planning this work the following scheme may be a useful starting point for the selection of the relevant factors to be studied.

### 2.3.1 Soil factors

a. *topography* ⎫these studies should give data on the degree of
b. *soil profile description*⎰gross variation in the soil under study.
c. *chemical analysis*: the degree of detail will depend on the requirements of the project; minimally it would probably involve assessment of total organic matter content.
d. *particle size analysis*: studies varying from simple mechanical analysis to the characterization of clay minerals and mineral components.
e. *pH*: the gross methods usually applied in these measurements give little or no data on the pH state at the microhabitat level.
f. *moisture status*: the determinations may vary from moisture content determinations (on an oven dry weight basis) on freshly taken soil samples to determinations of soil moisture characteristics (pF curves).
g. *temperature, $CO_2$ and $O_2$ status*: such measurements are frequently made on a continuous basis (using automated apparatus) over the whole experimental period.

### 2.3.2. Biotic factors

a. *Above ground*
    (i) Description of the pattern of the higher plant community (communities).
    (ii) Dominant and associated green plant species: determined at each of the seasons.
        In the course of (i) and (ii) the peculiarities in the vegetation and its distribution should be noted.
    (iii) Vegetational history (if possible): previous cultivation, burning or successional stages.

(iv) Animal effects: e.g., grazing.

(v) Input of organic matter: i.e., tree litter fall and production of ground vegetation, animal parts and faeces.

b. *Below ground*

(i) Root distribution in soil profile.

(ii) Movement of soil animals.

(iii) Input of organic matter: root production, soil animals (faeces, bodies, etc.).

### 2.3.3 Meteorological factors

a. *Solar radiation*: essential for studies on energy flow and productivity.

b. *Rainfall*

c. *Temperature*

It may be desirable to consider meteorological factors at both macro- and micro-levels.

No reference has been made to the procedural details for obtaining habitat data. Reference to standard works will provide the required data, e.g., Black *et al.* (1965); Metson (1956); Piper (1944); Newbould (1967).

B

# 3

# Sampling

## 3.1  Collection of samples from the field

Samples may be collected from the field for one of two main reasons. They may be required only as a source of micro-organisms, e.g. for isolation of antibiotic producers, or they may be needed for studies of the natural state of the soil.

In the former case, very few sampling precautions are necessary for in most cases it will not matter whether the sample has been disturbed or altered so as to cause changes in the microflora, providing that these changes have not led to wholesale elimination of all or part of the microflora. Even soil returned to the laboratories, sieved and air dried, and stored for some months, may be useful in such studies. The method described by Clark (1965) is suitable for this type of study. Here a thoroughly mixed gross sample is taken and a large amount, e.g. 2 lb, is placed in a polyethylene bag or a sealed, waxed cardboard container. This may be taken to the laboratory, only taking care not to expose it to undue heating or drying. The sample may be used at once, or stored at 4°C for up to two weeks. Prior to use, the sample may be passed through a 10-mesh sieve to homogenize the material.

However, if conclusions concerning the native environment are to be drawn from the results, it is usually necessary to try to preserve the sample in such a way that it is in the same physical, chemical and biological state when it is used as it was when it was taken from the soil. Complete stability of the sample is not likely to be achieved but one should be aware of the major changes which can occur, and it is clear that the method outlined above will not be suitable.

The most obvious change that can take place is a change in temperature. Air temperature and laboratory air temperature are frequently very different from soil temperature. Dark soils exposed to the sun may be considerably hotter than the atmosphere while frozen soils will be colder than ordinary laboratory temperatures. Two approaches have been made to this particular problem. Some workers prefer to refrigerate their samples, arguing that multiplication of cells will be slowed down. While this is undoubtedly true there have been very few studies on the effect of low temperatures on the death rate of soil micro-organisms and so we cannot be sure that the microflora remains stable. Other workers attempt to keep their samples at the temperature at which they were sampled, which is most easily achieved by keeping them in some sort of thermos flask. This is probably most satisfactory unless some other factor, which was limiting activity, has become non-limiting in the sample, e.g. aeration, in which case substantial and unnatural growth might result.

Amongst the other factors which are most likely to change in the sample are aeration and water content, with the most marked changes occurring near the surface of the sample. For this reason, it is often advisable to take a large soil block back to the laboratory and sample it again immediately prior to use. Even large blocks may change, however, especially if one is dealing with waterlogged soils for the excess water may easily drain away into the bottom of the soil container. In this case it may be best to take relatively small samples and keep them in plastic bags. The nature of the bags may be changed to suit the sample. If it is aerobic, then polyethylene (Stotzky *et al.*, 1962) should be used, which allows for passage of air but not water vapour. If the samples are anaerobic, impervious plastics or glass will be suitable.

Biological changes could also result from the contamination of specimens. Contamination may be from the air, from the sampling tool or soil container, or from soil one wishes to exclude from the sample. Contamination from the air is likely to be of little importance since its microbial content is very much smaller than soil. Its effect may also be countered by sub-sampling a large soil block in the laboratory, using strict aseptic technique. Contamination from sampling devices or soil might be more serious, but once again can be countered by sub-sampling if it is felt that gross contamination has occurred. In some cases, such contamination might be difficult to detect, i.e., the accidental mixing of soil on an exposed profile by careless digging, so care must be taken even before the removal of the sample.

The importance of all the changes dealt with so far may be aggravated or minimized by the duration of the storage period. It is likely that the average generation time of members of the soil microflora is long, except in soils with freshly added, readily available nutrients, e.g., rhizosphere soils and some leaf litters, so that even if a change takes place, it will be small if the sample is used soon after sampling. No accurate information on the permissible storage time for different soils is available and most workers are rather pragmatic in their outlook on this question. Since most sampling sites are within half a day's journey of the laboratory, samples are rarely kept for more than 24 hours. Until more reliable information is available, it would certainly be unwise to prolong this period and would be preferable to shorten it to a matter of minutes or hours, or eliminate it altogether.

A procedure that is frequently recommended prior to use of soil samples is sieving and mixing. In order to sieve soil, it may be necessary to partially dry it, thus causing changes in the sample.

Sieving and mixing of soil may well be necessary if one wishes to obtain a set of relatively uniform and reproducible samples, e.g. for testing the reliability of a particular technique, but it should be clearly recognized that it will provide a biased sample. Large particles may be excluded from the sample if they cannot be broken into smaller ones, e.g. coarse gravel, root and litter fragments, twigs, etc. Consequently, if it is desired to investigate the whole population, further sampling of the rejected fraction will be necessary.

The devices used to take samples from the field or from soil in the laboratory are very varied. Since their design is dependent upon the nature of the soil, i.e. consistency, depth, etc., and the investigation, the individual worker will have to design his own sampling tools. They may range from spatulas and spades to various coring devices including glass tubes, simple borers and complex mechanical augurs containing replaceable sterile sleeves for holding the sample.

## 3.2  Sampling units

It is important that experiments using natural soil are based on a sound sampling procedure. When soil is sampled and one or more of its properties are analysed, a number of errors influence the accuracy of the final result. These are (i) sampling error, (ii) selection error, and (iii) measurement error. As soil is so variable, (i) and (ii) often exceed (iii), so it is important that they are reduced as far as is practically and economically feasible.

In any study a basic *soil unit* must be defined. This can range from an acre field to a single horizon or particular type of soil particle, but whatever its nature, the results obtained will be related to it. Comparisons will be made between the properties of different soil units of equivalent level, e.g. between fields treated with different fertilizers or between organic and inorganic particles. However, the *samples* taken from a soil unit constitute only a small part of the whole unit but are assumed to be representative of it. Therefore, the parameters of the soil unit which are studied with the samples taken are not in fact *measured* but *estimated*. The accuracy of this estimate depends on (i) the variability within the soil unit, (ii) the number of samples taken and (iii) the way in which the samples are selected. It is thus of fundamental importance to assess the variability and reproducibility of samples taken to represent the same soil unit.

To do this it is essential to devise a plan of sampling which will allow statistical analysis of the results obtained. Selection of appropriate statistical procedures will largely depend on the problem under investigation and the advice of a statistician should be sought both on this and the planning of the sampling procedure.

In this section, a few of the generally useful statistical procedures are given. This is by no means a comprehensive treatment of statistics and serves only to show the relevance of statistics to problems of soil sampling.

### 3.3    Statistical procedures

#### 3.3.1    Types of distribution
Before deciding which type of analysis is required, it is necessary to find the type of distribution to which the data approximate.

The type often encountered is the *normal distribution,* in which the distribution is symmetrical and clusters around a central average value. This occurs when the variable measured is continuous with no limit to the number of individuals with different measurements, e.g. yields of a crop from a range of standardized plots.

Other types of distribution occur when the variables are discrete or discontinuous. If the incidence of some event is observed in sub-samples of uniform size, the sub-samples may be classified according to the numbers of times the event occurs. If there are two alternative characters which every individual must have, e.g. survival or death, a *binomial distribution* of the resulting frequencies arises.

When a particular event is rather uncommon but a sufficient number of instances are examined for several cases to be observed, a *Poisson distribution* occurs. Thus if cells are counted with a haemocytometer slide, although there is a very small chance of any one cell being in a given square, the total number of cells is so large that many squares will have one or more cells in them.

The assessment of the type of distribution encountered in nature can be complex and transformation of the data may make them more susceptible to analysis, e.g. the square root transformation for Poisson distribution; information on this problem is given by Taylor (1961), Cassie (1962) and Anthony (1970).

If there is evidence that sampled populations are so far from normal as to invalidate the methods of analysis, various *non-parametric tests* may be applied (Snedecor, 1962).

The methods of analysis given here apply to normally distributed data.

### 3.3.2 Assessment of the variability and reproducibility of samples

A measure of the variability of samples representing the same soil unit is given by their *standard deviation (s)*.

$$s = \sqrt{\frac{\Sigma(x-\bar{x})^2}{n-1}} \tag{1}$$

where $\Sigma$ = sum of

$x - \bar{x}$ = deviations from the mean $(\bar{x})$

$n$ = number of samples

$n-1$ = degrees of freedom.

The variance $(s^2)$, which it is often more convenient to use, is the square of the standard deviation.

$$s^2 = \frac{\Sigma(x-\bar{x})^2}{n-1} \tag{2}$$

Calculation of $(x-\bar{x})^2$ is laborious and it is quicker to use

$\Sigma x^2 - \dfrac{(\Sigma x)^2}{n}$ to calculate this, i.e. $s^2 = \dfrac{\Sigma x^2 - (\Sigma x)^2/n}{n-1}$

The standard error $(s_{\bar{x}})$ is a measure of reproducibility and can be calculated from the standard deviation or the variance.

$$s_{\bar{x}} = \frac{s}{\sqrt{n}} \text{ or } s_{\bar{x}} = \sqrt{\frac{s^2}{n}} \tag{3 and 4}$$

If the distribution of the estimate is approximately normal and the number of samples is greater than 30, the interval $\bar{x} \pm s_{\bar{x}}$ includes approximately two-thirds of the means of similarly drawn batches of samples, i.e. the range in which the means of future sample batches can be expected to fall. Standard errors can also be used to calculate confidence limits ($L$), which can be placed around the mean at a chosen level of probability. If the number of samples ($n$) exceeds 30, the confidence limits can be found from

$$L = \bar{x} \pm \frac{ds}{\sqrt{n}} \tag{5}$$

where $d$ is chosen from tables (Fisher & Yates, 1963) to correspond with the required probability level. If the number of samples is less than 30, the confidence limits can be found from

$$L = \bar{x} \pm \frac{ts}{\sqrt{n}} \tag{6}$$

where $t$, with $n-1$ degrees of freedom, is chosen from tables (Fisher & Yates, 1963) to correspond with the required probability level.

An estimate of the number of samples needed in future to give a selected confidence limit at a given level of probability can be obtained from

$$n = \left(\frac{ts}{D}\right)^2 \tag{7}$$

where $n$ = number of samples needed

$D$ = selected confidence limit.

### 3.3.3 Comparison of the means of two batches of samples

From information on the variability of each batch of samples, calculated separately, their means can be compared to see how significantly they differ.

Let us suppose that there were $n_1$ samples in the first batch, with a mean of $\bar{x}_1$ and a standard deviation of $s_1$; in the second batch the corresponding values were $n_2$, $\bar{x}_2$ and $s_2$.

If the batches consisted of more than 30 samples, their means could be compared by

$$d = \frac{\bar{x}_1 - \bar{x}_2}{\sqrt{\dfrac{s_1^2}{n_1} + \dfrac{s_2^2}{n_2}}} \tag{8}$$

If the calculated value of $d$ exceeds that of the value of table $d$ at the chosen probability level, the means are significantly different at that level. Thus a value of $d > 1 \cdot 960$ indicates significant difference at the $0 \cdot 05$ probability level and a value $> 2 \cdot 576$ indicates this at the $0 \cdot 01$ level.

If the batches consisted of less than 30 samples a different procedure is required.

The method used depends on whether or not it can be assumed that the unknown variances of the parameter throughout the units from which samples are drawn are the same or not. This can be assessed by a variance ratio test on the calculated variances $s_1^2$ and $s_2^2$ of the two batches of samples.

$$\text{Calculate } F = \frac{s_1^2}{s_2^2} \text{ (with } s_1^2 \text{ being the larger)} \tag{9}$$

Refer to variance ratio tables (Fisher & Yates, 1963) to find the value for $F$ at the required significance level corresponding to $n_1 - 1$ degrees of freedom in the numerator and $n_2 - 1$ in the denominator. If the calculated $F$ value exceeds that of the table value, the variances are significantly different at the chosen level.

If the variances are found to be not different, the means can be compared by

$$t = \frac{\bar{x}_1 - \bar{x}_2}{s\sqrt{\dfrac{1}{n_1} + \dfrac{1}{n_2}}} \tag{10}$$

$$\text{where } s^2 = \frac{\Sigma_1 (x - \bar{x}_1)^2 + \Sigma_2 (x - \bar{x}_2)^2}{n_1 + n_2 - 2}$$

The calculated $t$ value is then compared with the table $t$ value for the chosen probability level and the relevant degrees of freedom ($n_1 + n_2 - 2$).

If the variances are found to be different, the means can be compared by calculating $d$ from equation (8) and then treating $d$ as Student's $t$ with $f$ degrees of freedom, $f$ being found from

$$f = \frac{1}{\dfrac{u^2}{n_1 - 1} + \dfrac{(1 - u)^2}{n_2 - 1}} \tag{11}$$

where

$$u = \frac{\dfrac{s_1^2}{n_1}}{\dfrac{s_1^2}{n_1} + \dfrac{s_2^2}{n_2}} \qquad (12)$$

### 3.3.4 Comparison of the means of samples from more than two soil units

If more than two means have to be compared, an *analysis of variance* is required. This can also be used in all cases where a $t$-test is used.

Consider a study in which four samples were taken from each of three horizons in a soil and counts of microbes per gram obtained.

| | Counts of microbes Horizons | | | |
|---|---|---|---|---|
| | A | B | C | |
| Samples | | | | |
| 1 | $a_1$ | $b_1$ | $c_1$ | |
| 2 | $a_2$ | $b_2$ | $c_2$ | |
| 3 | $a_3$ | $b_3$ | $c_3$ | |
| 4 | $a_4$ | $b_4$ | $c_4$ | |
| totals | $a_r$ | $b_r$ | $c_r$ | Grand total $a_r + b_r + c_r$ |

$$\text{overall mean } (\bar{x}) = \frac{a_r + b_r + c_r}{12}$$

The mean of all 12 samples ($\bar{x}$) will have variance due to horizon difference ($H$) and also due to random difference between samples in the same horizon.

The total variance ($T$) can be estimated from the total sum of squares:

$$T = \text{ sum of squares of all deviations from the mean.}$$

This is found most easily from $\sum x^2 - \dfrac{(\sum x)^2}{n}$

$$\sum x^2 = a_1^2 + a_2^2 \ldots + c_3^2 + c_4^2$$

$$(\sum x)^2 = (a_r + b_r + c_r)^2$$

$$n = 12$$

$$T = a_1^2 + a_2^2 \ldots c_3^2 + c_4^2 - \frac{(a_r + b_r + c_r)^2}{12}$$

The variance due to horizon differences $(H)$ can be estimated from:

$$H = \sum \frac{\text{Horizon totals}^2}{\text{No. samples from horizon}} - \frac{(\sum x)^2}{n}$$

$$H = \frac{a_r^2}{4} + \frac{b_r^2}{4} + \frac{c_r^2}{4} - \frac{(a_r + b_r + c_r)^2}{12}$$

Analysis of variance can then be made thus:

| Source of variation | Sum of squares | Degrees of freedom | Mean square (variance) |
|---|:---:|:---:|:---:|
| Among horizons | $H$ | $3 - 1 = 2$ | $\dfrac{H}{2}$ |
| Between samples (residual) | $T - H$ | $11 - 2 = 9$ | $\dfrac{T - H}{9}$ |
| Total | $T$ | $12 - 1 = 11$ | $\dfrac{T}{11}$ |

It is then possible to decide if the variance due to horizons is significantly greater than that due to residual variation between samples.

A variance ratio $(F)$ is calculated:

$$F = \frac{\text{larger variance}}{\text{smaller variance}} \quad \text{i.e.} \quad \frac{H/2}{T - H/9}$$

The table $F$ value for a chosen probability and, in this case, 2 and 9 degrees of freedom is then found. If the calculated $F$ values exceed the table $F$, the variances are significantly different.

The variance between samples can also be used to estimate standard error and confidence limits for the whole set of *samples*.

It must be noted that although significant differences due to horizon may be found, it does not automatically follow that every horizon is different from every other.

When the nature of the experiment allows selection of plots of soil for sampling, a *randomized block design* is often useful. Consider an experiment to assess the yields of four different varieties of wheat. Each is planted in a number of plots and the yields from each plot measured. If soil conditions between the plots vary, this will contribute to the variation in yields of plots with the same variety and possibly obscure differences between varieties. It is therefore important to assess these two sources of variation.

The land is divided into say five blocks and each block divided into four plots, with each variety allocated randomly to one of the plots in each block. Soil conditions within any block should be fairly uniform, although large differences may occur between blocks. Results can be subjected to an analysis of variance in order to separate variety variance from block, i.e. soil, variance.

| Source of variation | Sum of squares | Degrees of freedom | | Mean square (variance) |
|---|---|---|---|---|
| Varieties | $T$ | $4-1$ | 3 | $\dfrac{T}{3}$ |
| Blocks | $B$ | $5-1$ | 4 | $\dfrac{B}{4}$ |
| Residual | $X-(TB)$ | $(4-1)(5-1)$ | 12 | $\dfrac{X-(TB)}{12}$ |
| Total | $X$ | $20-1$ | 19 | $\dfrac{X}{19}$ |

## 3.4 General Principles of Sampling

### 3.4.1 Planning

A number of important points should be borne in mind when planning a sampling programme.

(i) At least two and preferably more samples must be drawn from the same soil unit and evaluated separately to allow any assessment of variability to be made. Several small samples, required for any statistical comparison, drawn from one unit yield more precise information than a single large sample.

(ii) The soil unit itself should be as homogeneous as possible. The subdivision of heterogeneous units into homogeneous ones is an important way of increasing sampling accuracy. For example, if two soils are being compared and each has two distinct horizons, the horizons should be sampled separately as four units.

(iii) The *sampling area* which contains units to be studied should be as homogeneous as possible. Thus if a forest soil is to be studied, an area within the forest where topography, plant cover, soil type, etc., are reasonably uniform should be selected.

(iv) An unbiased estimate of the mean of samples drawn from the same unit requires that every sample has an equal independent chance of being drawn.

(v) An unbiased estimate of significance of the means of two or more batches of samples requires that every possible batch has an equal chance of being drawn.

### 3.4.2  Sampling procedures
A number of procedures which meet, at least partially, these requirements can be followed.

#### (a)  *Simple random samples*
If samples can be selected at random, requirements (iv) and (v) are met. This can be done by selecting sampling points within a unit using pairs of random numbers from tables (Fisher & Yates, 1963) to select distances on coordinates of a grid system laid out over the entire unit. The sample is taken at the point where the lines from the two coordinates intersect.

Variation among random samples taken from the same unit and also between units, can be estimated from variance (2) and standard error (3 and 4). Confidence limits (5 and 6) can be found and used to predict the number of samples required in future work (7).

#### (b)  *Stratified random samples*
These meet requirement (ii) when homogeneous sub-units are discernible within the unit. Stratified random sampling involves drawing samples randomly (as for simple random samples) from each sub-unit. In general, the more stratification (sub-division) which takes place, the greater will be the precision, but the extra time and effort involved must be weighed against this. The gain in precision depends on the variation between the stratum (sub-unit) means: the greater the differences, the greater is the increase in precision when they are treated separately.

Variance, standard error and confidence limits together with significance assessment can be obtained using the analysis of variance procedure (13). Confidence limits about the overall mean can be estimated using the between-sample variance in formulae 5 or 6.

#### (c)  *Systematic sampling*
In this procedure, samples are taken at regular distances away from each other in one or two dimensions. Usually parallel lines or grids are used to define the sampling points.

Systematic samples are useful except where there are periodic fluctuations within the unit being sampled, e.g. rows of plants in a cultivated field. If the sampling interval is equal to an integral multiple of the period in the unit, sampling will be highly selective, e.g., only soil under plants would be taken from the field.

Estimating the variability and reproducibility of systematic samples can be done in a number of ways:

(1) The samples can be assumed to be random and treated as simple random samples. This is dangerous and independents of samples should be tested.

(2) Discernible sub-units can be treated separately as in stratified random samples.

(3) A number of sets of separate systematic samples can be drawn at random, the sample means for each being calculated. The means are then treated as for a group of simple random samples equal in size to the number of sets.

## (d) *Sub-sampling*

Sometimes a sample is taken and then a number of smaller sub-samples are drawn from it at random. This has practical advantages, for an estimation of the properties of a large sample is obtained without dealing with the entire sample. However, it usually decreases the precision of the estimate, as an additional source of variation, i.e., that among sub-samples, is added to the sampling error.

Variation among sub-samples drawn from one sample can be estimated as for simple random samples. If more than one sample is sub-sampled, variance includes both that among samples and among sub-samples. To distinguish between these, analysis of variance (13) must be carried out as for stratified random samples.

## (e) *Composite samples*

In this procedure, samples representing a soil unit are drawn separately but then bulked and mixed. The whole mixture or sub-samples from it are then analysed. Compositing will have some validity but only under certain conditions and these are as follows:

(1) Equal amounts from each sample must eventually contribute to the entity analysed.

(2) No interactions affecting results should occur between mixed samples.

(3) The only objective of the study is an unbiased estimate of the mean; no estimate of precision is obtained.

(4) Analysis of two or more sub-samples from the composite cannot provide an estimate of variance among the samples as taken from the soil unit. This will, however, allow for comparison *between similar* composite samples.

The accuracy of the result obtained with a composite sample depends on the variability among the samples used to make it and the number of samples bulked. If an estimation of this variability can be obtained (by analysing samples separately and treating them as simple random samples), it can be used to determine the number of units to include in the composite to attain a given precision. This can be done using formula 7. If no estimate of variability of samples from the same unit is available, compositing should be avoided.

For more detailed information, the following should be consulted: Bailey (1959), Cline (1944), and Peterson & Calvin (1965).

# 4

# Determination of the Form and Arrangement of Micro-organisms in Soil

## 4.1 Introduction

Much valuable information on the form, location and growth pattern of micro-organisms can be obtained by examining the soil microscopically. For examination of the form of organisms growing in the soil, it is desirable to use high resolution microscopes and to attempt to remove the organisms from the confusing soil background. For examining location on particular soil particles and growth patterns through the soil, lower resolution techniques are useful in which it is essential not to remove the organisms from the soil.

There are several ways of categorizing growth patterns of micro-organisms, but in the context of the methods discussed here it may be useful to outline some of the possible patterns which can be detected. The terminology adopted by Gray & Williams (1971) is used here:

(i) Unicellular, restricted, non-motile pattern, e.g., some bacterial colonies, yeast cells; may be subdivided in regard to size and cell shape.

(ii) Unicellular, motile pattern, e.g., diatoms, some bacteria, chytrids.

(iii) Restricted mycelial pattern, e.g., *Penicillium, Streptomyces,* etc. where hyphae do not extend into the soil surrounding the colonized particle.

(iv) Locally spreading mycelial pattern, e.g., *Mucor ramannianus* which may spread into the surrounding soil where it forms a zone of chlamydospores.

(v) Diffuse spreading mycelial pattern, e.g., *Zygorhynchus* where isolated hyphae pass through the soil, unassociated with particular substrates.

(vi) Mycelial strand or rhizomorph pattern, e.g., many basidiomycetes which migrate from one substrate to another by means of well differentiated structures (mycelial strand or rhizomorph).

(vii) Plasmodial pattern, e.g., myxomycetes and myxobacteria, in which cells may aggregate and swarm over the soil surface.

*Chapter 4*

## 4.2 Direct examination of soil particles

### 4.2.1 Light microscopy

Soils frequently contain quite large organic and inorganic particles, e.g., humus, sand grains, etc. A good deal can be discovered concerning the arrangement of microbial cells in these soils by examining them carefully under the microscope. This is most useful when the organisms are small and in the form of restricted colonies, and less useful when large mycelial and spreading organisms are present which pass from one particle to another (see later for a description of the use of soil sectioning methods for studying such organisms).

Unstained particles may be examined by placing a few soil particles in a drop of water and examining them with or without the addition of a cover slip. It is often useful to examine small fragments of decaying leaves in this way, sometimes after incubation in a damp chamber, in order to reveal the density and occurrence of sporing and vegetative hyphae.

When soil particles in any given preparation are of differing sizes, it may be difficult to observe them because of the limited depth of focus of the light microscope. Indeed, the upper surface of big particles may be raised above the upper surface of small ones, making it impossible to focus on the latter without grinding the lens into the slide. These difficulties may be partly overcome by placing the particles to be examined on a bed of agar slightly thicker than the largest particle and pressing the particles into the agar with a cover slip (see Fig. 1).

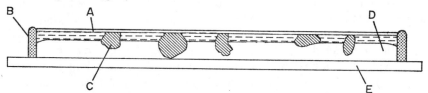

Figure 1. Method used for examining the occurrence of bacteria on sand grains. (A) cover slip; (B) varnish seal; (C) sand grain; (E) microscope slide.   From Gray T.R.G., *et al.* (1968) in *The Ecology of Soil Bacteria*, ed. T. R. G. Gray & D. Parkinson. Liverpool University Press.

Many organisms are difficult to see on soil particles unless they are stained, but if stains are used, it should be remembered that they may dislodge or displace micro-organisms. Soil can be placed in a glass cylinder fitted with a nylon weave base which is then lowered into a trough containing an appropriate stain (see Fig. 2).

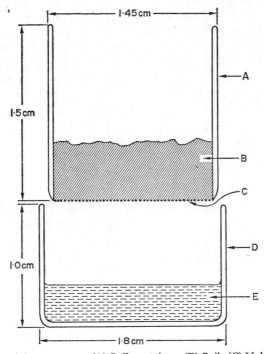

Figure 2. Soil staining apparatus (A) Soil container; (B) Soil; (C) Nylon weave base; (D) Staining trough; (E) Staining solution. From Hill, I.R. & Gray T.R.G. (1967) *J. Bact.*, **93**, 1888.

Among the most useful stains are the following:

(a) *Phenol-aniline blue* (phenol, 5% aqueous, 15 ml; aniline blue, 6% aqueous, 1 ml; glacial acetic acid, 4 ml). Filter 1 hour after preparation; stain for 1 hour, followed by rinsing in water and/or dehydration in 95% alcohol if required.

This is a good general purpose microbial stain that gives a fair contrast between cells and their background.

(b) *Acridine orange* (0.2%, or less, acridine orange in water used at a rate of about 0.6 ml/g soil for 3 minutes at room temperature followed by rinsing with 1.0% sodium pyrophosphate for 30 seconds).

This stain was originally used to distinguish between living and dead cells. Dead cells, lacking a cytoplasmic membrane soak up high concentrations of the stain and appear red when irradiated with ultra-violet light. Living cells largely exclude the dye and appear green. Unfortunately, the outcome of this stain is dependent on the concentration of the dye solution, the nature of the microbial cell wall as well as the degree of cell viability so it is now rarely used for this purpose. Recently (Graham & Caiger, 1969) have suggested the use of primulin (0·01 % for 3 minutes) which only stains dead cells, but this method has not been tested widely on organisms other than yeasts.

Acridine orange does have other uses. Many soil particles are opaque and can only be examined with incident light sources. Ordinary stains are not of much value in such situations, but fluorescent dyes are, since they give a high degree of contrast between cell and soil particle when irradiated with ultra-violet or short wavelength blue light. Since the majority of bacterial cells are closely associated with opaque organic material incident ultra-violet microscopy is of great value.

(c) *Fluoroscein isothiocyanate* (Pital *et al.*, 1966; Babiuk & Paul, 1970) (1·3 ml of 0·5M carbonate-bicarbonate buffer (pH 7·2); 5·7 ml of 0·85% physiological saline; 5·3 mg crystalline fluorescein isothiocyanate). Mix for 10 minutes; stain for 1 minute at 37°C, followed by rinsing for 10 minutes in 0·5M carbonate-bicarbonate buffer (pH 9·6) and mounting in buffered glycerol (pH 9·6).

This dye reacts specifically with proteins and is therefore useful as a general biological stain. The intensity is less dependent upon concentration and is more uniform from one organism to another, although the presence of pigments in hyphal walls may mask fluorescence to some extent (see also counting procedures, p. 64).

(d) *Fluorescent antisera*. The stains described thus far are non-specific and will react with most or all micro-organisms. Only in a few cases is it possible to recognize distinct species or genera of micro-organisms prepared with or without these stains, e.g., the typical branching pattern of *Rhizoctonia* mycelium (Blair, 1945). However, it is possible to prepare antisera that will react specifically with species or even strains of organisms. This reaction can be used to identify organisms in field soils under the microscope if fluorescent dyes are coupled to the antisera (see Fig. 3). Fluorescent antisera for a number of soil organisms have been produced and the procedures used for two of these, a bacterium and a fungus, are given:

(i) Production of fluorescent antisera against vegetative cells of *Bacillus subtilis* (Hill & Gray, 1967). *Bacillus subtilis* cells are grown in agitated nutrient broth or soil extract at 25° for 12–18 hours. The cells are centrifuged and washed six times in physiological saline. The cells are resuspended in saline and the turbidity of the suspension matched with a Brown's opacity tube

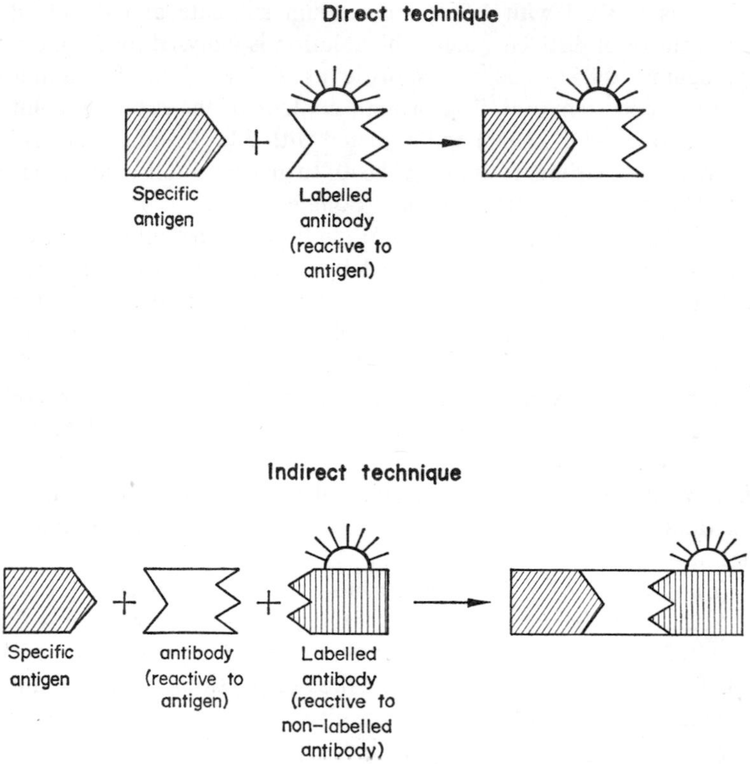

**Direct technique**

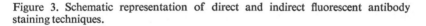

**Indirect technique**

Figure 3. Schematic representation of direct and indirect fluorescent antibody staining techniques.

(No. 8). 'O' antigens are prepared by heating these suspensions at 100° for 60 minutes, cooled and stored at 4°. The antigens are injected intravenously into rabbits as follows: day 1, 0·5 ml; day 4, 1·0 ml; day 7, 2·0 ml; day 10, 3·0 ml; day 13, 3·0 ml; day 16, 3·0 ml; day 19, 3·0 ml; day 22, 3·0 ml. On days 32, 33, 36 and 37, the rabbits are bled from the marginal ear veins. Secondary injections and bleedings are carried out after a 1-month rest period.

The blood samples are allowed to clot and the clear supernatant collected and sterilized by membrane filtration for prolonged storage.

The $\gamma$ globulin fraction of the antisera may be precipitated by mixing equal volumes of antiserum and 3·9M ammonium sulphate at 2°. The resulting precipitate is washed with 1·95M ammonium sulphate and dissolved in a minimal volume of distilled water. The solution is dialysed (in 2 cm diameter tubing) against 0·85% sodium chloride at 2°C until all the ammonium sulphate has been removed. The protein content of the resulting solution is estimated (Lowry *et al.*, 1951) and adjusted with 0·85% sodium chloride and 0·5M carbonate-bicarbonate buffer (pH 9·0) to give a sample containing 10 mg of protein/ml and 10% carbonate-bicarbonate buffer.

To conjugate this preparation, crystalline fluorescein isothiocyanate (BBL) is added over a 1-hour period, followed by continuous stirring for 12 hours at 2°. Permissible dye : protein ratios are 1 : 40 to 1 : 100. The conjugated sample is then passed through a Sephadex (G25 bead form) column (25 × 2·5 cm) equilibrated with 0·01M, pH 7·1 phosphate buffered saline. The conjugated dye passes rapidly through the column and may be collected (free from uncombined dye) after about 15 minutes, in small bottles and stored at −20°.

It is advisable to check the specificity of the conjugate by attempting to stain a range of different bacterial species, before using it to stain mixed populations in soil.

(ii) Production of fluorescent antisera against hyphae of *Aspergillus flavus* (Schmidt & Bankole, 1963). *Aspergillus flavus* is grown for 4 days in shake culture in a glucose, dicalcium glutamate, inorganic salts medium at pH 7·3. The mycelium is filtered off, washed in physiological saline and then homogenized in a blendor in saline to give a concentration of 0·1 mg dry weight mycelium per milliliter saline. This suspension is injected intravenously into rabbits as follows: day 1, 0·5 ml; day 2, 1·0 ml; day 4, 1·0 ml; day 8, 1·5 ml; day 10, 2·0 ml; day 12, 2·0 ml; Subcutaneous injections of 2·0 ml are made on the 16th day and 18th day and the blood is sampled on the 25th and if necessary on the 27th, 29th, and 31st days. Agglutination titres should be at least 1280 at this time.

The procedure for preparation of fluorescent antisera is then the same as that previously described for *Bacillus subtilis*.

(iii) Staining of samples with fluorescent antisera: Fluorescent antisera may be used to stain soil particles and sand grains as already described. Rossi–Cholodny slides may also be stained. Soil particles are stained with antiserum for 30 minutes, allowing about 0·06 ml of conjugate for each 0·1 g of soil and agitating the particles in the stain at 10-minute intervals (see p. 21). They should then be rinsed in 0·01M, pH 7·1 phosphate buffered saline 2 or 3 times and placed in a small drop of water on a slide.

Slides may be stained in a similar manner, using only a small amount of conjugate. It is necessary to keep the slide in a damp chamber to prevent drying out.

A problem frequently encountered, especially with soils containing much clay, is the non-specific adsorption of dye by the soil particles. The following technique described by Bohlool & Schmidt (1968) helps to overcome this problem. It depends on pretreating slides with a gelatin-rhodamine conjugate which prevents non-specific staining and acts as a counterstain for the background. A 2% aqueous solution of gelatin (Difco Labs) is adjusted to pH 10–11 with 1N NaOH and autoclaved for 10 minutes at 121°; the autoclaved solution is readjusted to the same pH. The gelatin is conjugated by dissolving rhodamine isothiocyanate (Nutritional Biochemicals Corp.) in a minimum amount of acetone and adding this to give 8 $\mu$g dye to each milligram of gelatin. The mixture is stirred gently overnight and then passed through a Sephadex gel (G25 coarse bead form) equilibrated with phosphate buffered saline at pH 7·2. The red conjugate is collected in about 15 minutes. It may be preserved at $-20°C$ with the addition of 1 : 10,000 merthiolate. About 0·5 ml conjugate is flooded across half of the surface of the air-dried, heat fixed preparation. The slide is placed in an oven at 60° until the gelatin has dried and then allowed to cool. The slide is then ready for staining with the fluorescent antiserum as previously described, although staining times of up to 45 minutes may be used.

(iv) Observation of stained preparations: Preparations should be observed under ultra-violet or short wave-length blue light. This may be provided by HBO 200 mercury vapour lamps fitted with Schott BG12 filters, using OG1 glasses as barrier filters. Transmitted light is passed through a dark field condenser and incident light through a vertical illuminator. If preparations are to be mounted and observed under oil immersion, it is essential to use a non-fluorescent immersion oil, e.g., Leitz. A high dry objective lens ($\times 70$) is often useful to overcome problems from the use of fluorescent mounting fluids.

Stained cells should fluoresce a bright apple green. The background should be non- or weakly fluorescent, unless a rhodamine-gelatin conjugate has been used when soil particles appear an orange-red colour and the background dark. This is the only technique that will allow identification of organisms in soil, without lengthy isolation and diagnostic procedures.

(e) *Fluorescent enzymes.* Millar & Casida (1970) suggest that lissamine rhodamine B–200 sulphonyl chloride labelled lysozyme may react specifically with mucopeptides in bacterial cell walls and therefore may be a useful bacterial stain. Rhodamine compounds can also be obtained as isothiocyanates (Nutritional Biochemicals Corporation) and conjugated with lysozyme in the way already described for other proteins (p. 24). The stain is used as follows.

Preparations are treated with a 1·5% solution of sodium hypochlorite for 3 minutes to expose the glycopeptide in walls of Gram-negative organisms and spores. The preparations are washed in phosphate buffered saline for 10 minutes, stained with the labelled lysozyme for 60 minutes and finally rinsed again in phosphate buffered saline. If necessary specimens may be treated with Bohlool & Schmidt's gelatin counterstain (using FITC) to prevent non-specific adsorption of the lysozyme to soil particles (see p. 25). Some fungi also stain with labelled lysozyme, presumably because they contain other amino sugar compounds susceptible to this enzyme.

Observations are made with ultra-violet light as outlined on p. 25.

### 4.2.2 Electron microscopy

Use of electron microscopes allows the examination of soil and soil micro-organisms at much higher resolutions than those obtained with the light microscope. Three techniques have been used and these are described in outline below:

(a) *Scanning electron microscopy* (*Gray, 1967*). With this technique it is possible to examine micro-organisms in place on the surfaces of soil particles since an image is formed by electrons liberated from the surface of the specimen, collected and passed through photomultipliers. The degree of resolution is always better than that obtained with the light microscope at equivalent magnifications. As a result, an almost three-dimensional effect is produced even on ordinary photographs.

A specimen holder, a small metal stub—diameter 12·5 mm—is coated with an adhesive. While still sticky, soil particles may be scattered on the holder, or it may be pressed gently on to a freshly exposed soil face. Locci & Quaroni (1969) used another method in which soil cores in tubes are placed in beakers, and Freeze-up Freon (Mann Research Labs, New York) sprayed into the beaker. The soil freezes in a few minutes, especially if a small amount of water is added to the soil. The tube contents are extruded and the soil column sectioned into discs with a blade. The discs are then stuck on to the specimen holders. After tapping the holder to remove loose particles, the specimen is coated under vacuum with metal, e.g., gold-palladium alloy, copper, aluminum, in order to minimize charging effects, preferably rotating the specimen during the coating procedure.

The specimen may then be examined with the scanning electron microscope. Fungal hyphae are easily detected at low magnifications, e.g., × 100–200, but bacteria and actinomycetes are more difficult to find. Particles should be searched carefully at magnifications of × 1000–5000. Some cells show signs of distortion because of the preparation and it may be necessary to devise suitable fixing or freeze-drying techniques if this proves troublesome.

(b) *Transmission electron microscopy-sections* (*Jenny & Grossenbacher, 1963*). To examine micro-organisms *in situ* with the transmission electron microscope requires the preparation of ultra-thin sections. So far, the technique has only been used to study the high concentrations of micro-organisms present in the root-soil boundary zones. The technique has the advantage that cells buried in mucilage, etc. can be seen, though these would be invisible in the scanning electron microscope.

To prepare root-soil systems for study under the transmission electron microscope Jenny & Grossenbacher grew barley seeds in tubes filled with bentonite clay and permutite sand. The water content was adjusted to field capacity and after a few days the spaces between the soil particles were filled with liquid monomer which hardened to a rock hard consistency. Sections (10–50 nm thick) were then cut from the tubes using special diamond saws and knives.

(c) *Transmission electron microscopy-individual cells* (*Nikitin, 1964*). Soil is shaken with water (1 : 2 v/v) and allowed to stand for a short time. If mineral particles are still suspended in the upper part of the liquid, they may be centrifuged out at 2000 rpm for 2 minutes. Drops of the liquid are placed on formvar coated grids and stained with 1 % phosphotungstic acid solution or shadowed with chromium.

Alternatively, the soil-water suspension may have agar added to it (2%), be sterilized at 1 atmosphere pressure and poured into dishes. Instead of heat sterilization, poured plates may be sterilized with ultra-violet radiation. The surface of the agar is inoculated with a soil-water suspension and cells from colonies which develop on the medium may be picked off and prepared for examination on formvar coated grids. Similar soil-water-broth cultures may be examined in the same way.

Cells of unusual morphology, not detected in other ways, are observed using these methods. However, great care must be taken in interpreting the structures seen in untreated soil suspensions, as many of them may not be micro-organisms but detached pieces of soil animals, algae and fungi.

### 4.3   Direct Examination of the soil profile

Exposed soil profiles may be examined directly in the field, using the techniques described by Kubiena (1938). These involve special incident light soil microscopes fixed to the soil profile or on special stands. The main problem in these techniques is obtaining concentrated illumination and high magnifications for both observation and photographic records. In general, these techniques have to be used in conjunction with many of the methods already described for examination of fresh soil in the laboratory. Some special applications include examination of the relationships existing between mycorrhizal fungi and the roots of higher plants, and the growth of sporing heads of fungi within the soil.

### 4.4   Soil sectioning

Soil sectioning allows observation of the micromorphology of soil (the soil being in an undisturbed state). It aims, when applied to studies in soil microbiology, to show relations of micro-organisms to soil microhabitats, and has been infrequently used for quantitative studies on soil fungi.

#### 4.4.1   Technique for soil (Burges & Nicholas, 1961)

Blocks of soil approximately 10 cm × 10 cm × 2 cm are collected (the size of the initial sample will depend on the horizon characteristics of the soil under study). Single blocks of soil approximately 2 cm × 1 cm × 1 cm are placed in liquid nitrogen to quick-freeze them, freeze-dried and then impregnated with resin. The samples can be fixed and stained prior to embedding, if necessary to detect hyaline hyphae. This procedure is described by Alexander & Jackson (1955). Burges and Nicholas (1961) found that the increased number of operations required for staining soil preparations lead to disturbance of the soil and to poorer preparations.

The resin mixture, e.g., Marco (SB 28/C) or Bakelite (SR 17451 or SR 17497) is placed in an appropriate plastic tray and the sample is added to the resin.

The proportions of the Marco components used by Burges & Nicholas

(1961) are: resin, 80 ml; monomer, 16 ml; catalyst paste, 1 gm; accelerator, 3 ml. Impregnation is carried out under reduced pressure. The Bakelite components are: resin, 100 ml; catalyst, 2 ml; and accelerator, 2 ml. The impregnated samples are removed from the moulds and the under surface of the sample is ground down and polished. The soil samples are ground on geological laps using carborundum powder, grade 150, followed by a fine abrasive (carborundum powder, grade 220) and finally the surface is ground on a ground-glass plate using carborundum powder grade 600. Periodic washing of the sample with water prevents scratching of the section surface. The sample is then fixed to a slide. If the Bakelite resin is used, the resin itself can be used to cement the sample to the slide for grinding and as a mountant. When Marco resin is used, Lakeside 70 may be used as a cement and Canada Balsam as a mountant. Burges & Nicholas (1961) found that a soil section thickness of 50 $\mu$m (determined by the use of a polarizing microscope) was convenient for the examination of hyphae.

### 4.4.2 Technique for leaf litter (Minderman, 1956)

Leaf litter samples are collected and kept at $-10°C$ for at least 12 hours. The sample is kept frozen and slowly immersed in a 5 % gelatin in water solution at 35°C, and then in 10, 15 and 20 % gelatin solutions. Each immersion takes 1 to 2 hours. The container of 20 % gelatin solution and soil samples is cooled after the impregnation process and the litter sections (cubes) are cut out of the excess gelatin and fixed in 10 % formalin. The formalin solution should be changed during a fixing period of at least seven days. The samples are then placed in 80–90 % methyl alcohol until the sections are hard enough to cut.

A microtome can be used to cut 8 to 10 $\mu$m sections. Preparation can be made temporary or permanent, stained or unstained. Minderman found that Johansen's quadruple staining method was best.

The soil sectioning technique under-estimates the amounts of mycelium present in soil samples but allows detailed examination of relatively undisturbed soil under high magnification. It is extremely difficult to see bacterial cells by this method.

If the soil sectioning technique is to be used to its best advantage in studies in microbial ecology it should be applied to problems requiring data on the patterns of growth of soil fungi and the association of these micro-organisms to the varied soil microhabitats.

### 4.5   Contact methods

These techniques involve contact between soil and glass or other materials. The period of contact may be only brief, the material being pressed against a soil surface so that microbial cells present adhere to it. Thus the natural growth of micro-organisms in soil is examined by these so-called impression techniques. Alternatively the artificial surface introduced into the soil may be left buried for some time, allowing the growth of microbes over it to be examined.

### 4.5.1   Buried slide technique

A method for the direct microscopic examination of soil microbes was described by Rossi *et al.* (1936). Heat sterilized microscopic slides are buried in soil and extracted after given times. Soil from one side is removed and after heat fixation, the slide is treated with an appropriate stain (e.g., phenolic aniline blue). If fluorescent microscopy can be used, various fluorescent stains (see p. 21) may be applied.

Microscopic examination reveals the nature and growth pattern of soil microbes growing on the slide surface. Positive identification of the microbes is not often possible, unless characteristic fruiting structures are produced. Specific microbes may be detected on slides if they are stained with an appropriate fluorescent antibody preparation (see p. 22). Measurements of the frequency and quantities of microbial cells may also be obtained.

It cannot be assumed that growth on the inert glass surface is similar to that in the undisturbed soil. There is evidence that the insertion of the slide, providing a continuous surface on which condensation occurs, stimulates microbial activity. Therefore, observations of microbial growth patterns and interactions on slides must be treated with caution. If results are expressed qualitatively, a simple and reasonably accurate comparative assessment of the microbes present in different soil systems is obtained. In addition to its simplicity, the method has the advantage of being applicable in field conditions.

The buried slide method may be modified by coating slides with a particular substance and following their colonization in comparison with uncoated control slides. Provision of a substrate often results in more prolific fruiting of many microbes, thus allowing more accurate identification. A number of substances have been used, but the most successful have been those which are obtainable as transparent films which can be fixed to the slides.

Tribe (1957) described a method for studying microbial growth on cellulose.

Pieces of boiled and washed Cellophane are damped with sterile water and placed on glass cover slips which are then buried in soil. On recovery, microbial material is stained with picronigrosin in lactophenol.

Similarly, strips of chitin may be attached to slides and buried in soil (Gray & Bell, 1963) or cutin (Gray & Lowe, 1967) (see also p. 53).

### 4.5.2 Impression techniques

(a) *Rossi* et al. (*1936*). Microscope slides, uncoated or coated with agar or gelatin, may be used to obtain impressions from exposed soil surfaces. The slides are pressed firmly against the soil, removed, fixed and stained. Some microbial cells adhere to the glass or coated surface but in sandy soils little is picked up.

(b) *Brown* (*1958*). Sterile glass slides are smeared with nitrocellulose in amyl acetate $+5\%$ castor oil. The adhesive is spread over the central area of the slide, immediately prior to it being pressed against an exposed soil face for 20 seconds. It is important to spread the adhesive thinly and evenly and to press lightly so that only material flush with the soil surface adheres. Excess material not in contact with the adhesive is removed by gently tapping the slide after air drying. The films are stained for 1 hour in phenolic aniline blue, rinsed in sterile distilled water and dried. Successive peelings from the same area of the soil face may be taken to build up a three-dimensional picture of microbial distribution. Preparations can be examined by refracted or transmitted light.

The method is quick and simple and shows spatial distribution of microbes in natural soil. It is particularly useful for detection of fungal hyphae which can be relatively easily observed on the films. Results may be expressed quantitatively, allowing comparison of amounts of material in different soil systems. Selectivity in picking microbes out of the soil is likely to occur as some (e.g., those growing within substrates) are likely to be missed. Also satisfactory films of soils with a high organic content are not easily obtained because of clumping of aggregates.

(c) *Lingappa & Lockwood* (*1963*). This method can be used to obtain impressions of soil in laboratory conditions. The soil is packed into a suitable container, e.g., a Petri dish, and its surface smoothed over with a sterile implement. Its water content is then adjusted to about $25\%$. Impressions are taken immediately or after a period of incubation, in the following way:

Two to three drops of aqueous phenolic rose bengal solution ($1\%$ rose bengal, $5\%$ phenol, and $0.01\%$ $CaCl_2$) are placed on the soil surface and allowed to spread into the soil. The solution stains microbial material but not soil particles.

One or more drops of collodion (1·5% pyroxylin in 1 : 1 (v/v) absolute ethanol and ethyl ether) are placed on the soil surface and allowed to spread and dry to a thin film. This is removed with forceps and placed in a drop of mineral oil on a glass slide and covered with a cover slip for microscopic examination of adhering microbial cells. Successive peelings of the same surface may be taken.

Other recovery films may be used, e.g., 1–2% polystyrene (plastic Petri dishes) in benzene–toluene (2 : 1); molten (42°) 3% water agar.

Preparations can be made from soil samples as they are brought in from the field. Alternatively, microbial development in sterile or non-sterile soil may be followed under laboratory conditions. It is most satisfactorily applied to fine-particled soils which provide the essential flat surface more easily. Often it is necessary to sieve soil before use, to remove larger aggregates. Some microbial cells are not recovered and others stain only weakly or not at all.

### 4.6   Capillary and nylon mesh techniques (Aristovskaya & Parinkina, 1962; Perfiliev & Gabe, 1969; Waid & Woodman, 1957; Nagel-de-Boois, 1970)

Techniques for studying the development of micro-organisms on glass slides have been criticized for reasons other than those stated above. First, glass slides present a continuous flat surface, unlike the system of pores existing in soil, and secondly micro-organisms colonizing slides may be disturbed when the slides are removed from the soil. In order to stimulate the soil pores more closely, so that the rate and nature of microbial colonization can be studied under more realistic conditions, various Russian workers have developed the so-called capillary techniques.

Small glass capillaries (pedoscopes), preferably with a square or oblong cross-section (Fig. 4) are buried in soil, arranging them in the direction of moisture flow. The insides of the capillaries may be coated by filling the capillaries with nutrients in solution, removing the excess liquid with absorbent cotton wool, and allowing them to dry. Capillaries can be buried in this form, or partially filled with water.

After burial, the capillaries are left to equilibrate for 1–3 months and then removed and stained with suitable dyes. Capillaries can be removed at regular intervals to examine temporal or spatial variations in the microflora.

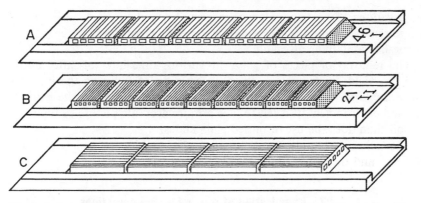

Figure 4. Arrangement of the pedoscope with different types of capillary cells. (A) Large size capillary cells with canals in cross section in relation to the axes of the adaptor. (B) Capillary cells with canals similarly situated, but considerably smaller in size. (C) Capillary cells with lengthwise orientated canals. Length of adaptor 70 mm. From Perfiliev B.V. & Gabe D.R. (1969) *Capillary Methods of Investigating Micro-organisms*. Oliver & Boyd, Edinburgh.

If they are examined without staining, they can be reburied with a minimum disturbance to the organisms inside the capillary.

Alternatively, instead of true capillaries, two microscope slides can be mounted together with suitable spacers to leave a slit of about 0·1 mm.

While these techniques are undoubtedly improvements on the contact slide methods, they have not been widely used because some effort is involved in making or acquiring suitable capillaries. For this reason many workers prefer to use the nylon mesh technique (Waid & Woodman, 1957) in which soil pores are simulated by the holes in close woven nylon material. A recent modification of this technique has been described by Nagel-de-Boois (1970).

Nylon gauze of appropriate mesh size provides a biologically inert matrix on and through which microbes can grow and which does not impede diffusion of liquids or gases or the growth of rootlets. Thus buried pieces of nylon gauze, after some time, come into equilibrium with the soil environment. The regular mesh pattern of the gauze facilitates observation of the amount and form of microbial development.

Pieces of nylon gauze ($10 \times 15$ cm, with 125 meshes per cm) are buried in soil and, at regular intervals, dictated by the aims of the project in hand, five pieces are removed from the soil and examined under the microscope for fungal development. The pieces of gauze can be stained to facilitate these observations. Two hundred meshes are observed on each piece of nylon gauze removed from the soil, and fungal growth is expressed as the percentage of meshes containing mycelium.

The removal of pieces of gauze at various time intervals after burial in soil has yielded information on the succession of fungi on the inert nylon substrate, and, because of the more natural equilibration of the gauze with the soil environment, it is felt to give a more accurate picture of fungal form in soil than does the use of buried glass slides.

The technique has also been used in studies on biomass and rate of growth of fungi and these uses are discussed on page 93.

## 4.7 Examination of root microbe interactions

### 4.7.1 The Fåhraeus slide technique (Fåhraeus, 1957)

Some micro-organisms, in addition to living in the soil, may infect plant roots and lead a mutualistic or parasitic existence. It may be necessary to observe the changes which take place on the root surface during the early stages of infection and a technique, devised by Fåhraeus, for examining interactions between root hairs and rhizobia may be used to do this. In this technique, germinated seeds are allowed to grow in a thin film of agar on a microscope slide, placed in a plant nutrient solution containing bacteria, allowing continuous observation of developing root hairs and bacteria without disturbance of the roots. It is only suited to small seedlings and great care has to be taken to avoid serious contamination during observation.

A nitrogen-free mineral solution is made up as follows: $CaCl_2$, $0.1$ g; $MgSO_4.7H_2O$, $0.12$ g; $KH_2PO_4$, $0.1$ g; $Na_2HPO_4.2H_2O$, $0.15$ g; Fe citrate, $0.005$ g; Mn, Cu, Zn, B, Mo, traces; distilled water, 1000 ml; pH $6.5$ after autoclaving. The solution is dispersed in 25 ml amounts in glass tubes ($c.$ $39 \times 125$ mm), plugged with cotton wool and sterilized at 15 lb/in$^2$ for 20 minutes.

Microscope slides ($26 \times 75$ mm) and cover slips ($24 \times 40$ mm) are dry sterilized in Petri dishes.

Seeds of the test plant are surface sterilized by successive treatment with $95\%$ ethanol and equal parts of $0.2\%$ formaldehyde and $0.2\%$ mercuric chloride, followed by thorough washing with several changes of distilled water. The washed seeds are then germinated in sterile water at room temperature and when 2 days old are transferred to the microscope slides as follows: Six or seven drops ($c.$ $0.2$ ml) of $0.3$–$0.4\%$ Bacto agar in the above mentioned mineral solution are pipetted on to half the slide and allowed to set. The seedlings are placed with their root tips in the agar and cover slip laid on top. If necessary the seed coat may be removed.

The completed slides are then placed in the tubes of nutrient solution and inoculated with a few drops of a thick suspension of bacteria, e.g., *R. trifolii*. Alternatively, the bacteria may be incorporated in the agar. The tubes are incubated at 25° in the light.

For observation, the slides are removed and excess solution is removed with sterile filter paper. The slides are examined with a phase contrast microscope, using glycerol (easily removed with water) if an immersion objective is used. After observation, the slide is replaced in the solution for a further incubation period.

### 4.7.2 Root observation boxes (Krasilnikov, 1958)

A number of methods have been developed in attempts to observe directly microbial development in rhizospheres, e.g., the application of Rossi–Cholodny slides; the growth of roots along specially mounted glass plates in soil; the use of observation boxes (miniature or large). These are outlined in the extensive review of rhizosphere microbiology by Krasilnikov (1958), and only one of the methods is described here.

A glass plate is mounted to form one easily removable large side of a box (5–6 cm wide, 20–25 cm high and 30 cm long). The box is filled with soil or sand in which plants are grown, and placed at a slope of 40–50° to ensure root growth along the glass plate.

Microscope slides are placed on the inner surface of the glass plate which can be removed after roots have grown over them. These slides can then be examined microscopically.

# 5

# Isolation of Micro-organisms

## 5.1 Introduction

The isolation of a soil microbe involves the transfer of its propagules from soil to artificial conditions conducive to growth, usually with the aim of obtaining a pure culture. This is an essential prerequisite to the study of the nature and activities of an individual microbe. However, soil is an extremely heterogeneous system, providing suitable microhabitats for a great variety of micro-organisms and because of the diversity of growth requirements of soil microbes, it is impossible to devise a single procedure by which all forms present in a soil sample can be isolated. Therefore, any isolation technique is to some extent selective and is designed to detect microbes with particular growth forms, biochemical capabilities or other desired properties. The choice of the most appropriate method is not always easy because of the multiplicity of techniques described. Therefore, in this section emphasis will be laid on the applicability of each technique described although it must be emphasized that, in many cases, small alterations to a method may considerably change its applicability.

The selectivity of an isolation procedure can be influenced at a number of stages. First, the method used to transfer microbial propagules from soil to artificial media is important. This may involve a deliberate selection and transfer of particular propagules, e.g., fungal hyphae, or an accidental selection, e.g., microbes most easily detached from their substrates. Secondly, pre-treatment of a soil sample before transfer to isolation media may eliminate or increase certain fractions of the microflora. For example, heating of a sample to 45°C will kill many bacteria but not affect thermotolerant propagules. Thirdly, selectivity can be drastically influenced by the artificial conditions into which the propagules are introduced. Various nutrients and microbial inhibitors can be incorporated in media and a variety of different incubation conditions used. The choice of these conditions may result in a large percentage of transferred propagules growing or, on the other hand, allow growth of only a single species.

Following isolation it is frequently necessary to establish whether the organism obtained is the one required, or a chance contaminant not involved in the process being studied. It is desirable, therefore, to apply Koch's postulates, originally designed for proving the involvement of organisms in disease production. These postulates require that:

(1) a particular organism always be found in association with the disease (or other phenomena);

(2) this organism be isolated and obtained in pure culture;

(3) on introducing this pure culture back into a healthy susceptible host (or sterile material) the disease (or phenomenon) be produced;

(4) the organism be re-isolated from the material produced in step 3.

In effect, these postulates are a statement of scientific procedure (Allen, 1959) but there are certain circumstances where they can be misleading. Thus, as Buddenhagen (1965) remarks, rigidly following Koch's postulates, combined with the use of selective isolation techniques, tends to narrow the recognition of micro-organisms, as secondary- or co-pathogens in disease.

## 5.2 Direct methods of isolation

These methods have been devised in order to isolate micro-organisms from soil in a particular phase of growth. In practice they have been used to isolate organisms present in soil in the vegetative state. Because of the size of the structures involved these techniques have been used mainly for the isolation of soil fungi, but direct sampling of capillary water, and subsequent isolation of bacteria therefrom has been attempted (Kubiena, 1938) as well as direct isolation of bacterial cells (Casida, 1962).

### 5.2.1 Direct isolation of fungi (Warcup, 1955)
This method is based on the observation that many fungal hyphae are associated with the heavier soil particles, i.e., those which are discarded in many soil dilution plate procedures.

A soil crumb (or small soil sample) is saturated with water and is then broken with a jet of water. The heavier soil particles are allowed to sediment whilst the finer particles are decanted off. This washing–decantation procedure is continued until a residue consisting only of the heavier soil particles remain. These residual particles are spread in a thin film of sterile water and examined with a stereo-dissecting microscope. Fungal hyphae are picked up with sterile instruments (fine forceps, needles) and plated on to a nutrient

D

medium (marking the position of the inoculum on the base of the plate). It is sometimes necessary to attempt removal of bacterial cells, other propagules and organic matter adhering to the hyphal fragments. This is done by carefully drawing each isolated hyphal fragment through semi-solid agar. The plates prepared are then incubated and the plated fragments are observed daily to ensure that any growth on the plates originates from the plated hyphae and not spores which have remained adhering to the hyphal fragments. Fungi growing on the plates should be subcultured on to slopes of nutrient medium and identified (see Chapter 8).

This method is tedious to apply, and is impractical for investigations which involve regular isolations from large numbers of soil samples. Considerable technical skill is required in the execution of the method. It is selective for hyphae of large diameter, hyphae which do not fragment easily, and dark-coloured hyphae, while hyphae which are closely associated with humus or organic matter fragments are frequently neglected.

Information on the killing of hyphae as a result of the isolation method is difficult to obtain but in many cases a large proportion of plated hyphal fragments fail to grow.

### 5.2.2 Direct isolation of bacteria (Casida, 1962)

A 1 g sample of soil is taken and 10 ml acridine orange solution added. The appropriate concentration, i.e., one which causes the bacterial cells to appear green, under ultra-violet illumination (usually between 1 : 1000 and 1 : 5000) should be determined by preliminary experiment. The mixture is shaken for 10 minutes.

A thin film of the mixture is spread on a slide which has been coated with a layer of agar. This slide may be observed directly under the ultra-violet microscope or may be incubated for 4 hours before observation, care being taken to prevent desiccation of the agar film. This pre-incubation period allows bacterial cells to reach the late lag growth phase when the cell size is at its largest.

Bacterial cells showing green fluorescence are transferred by micro-manipulation into a nutrient medium.

This technique demands expensive equipment and considerable technical skill. It is time consuming in application and therefore is not feasible for investigation involving regular isolation from many soil samples.

The advantage of this technique is that it allows correlation of the morphology of bacterial cells in pure culture with that of the cells in soil.

### 5.2.3 Other direct methods of isolation

Other methods have been described as direct methods, including: the immersion tube technique (Chesters, 1940) and its derivatives (e.g., Thornton, 1952, 1958; Parkinson, 1957) for the isolation of fungi. In these methods the isolation medium is inserted in the soil but is separated from the soil by an air gap. Active growth of fungi must occur for an organism to colonize the isolation medium.

Flotation methods for *Helminthosporium sativum* spores (Ledingham & Chinn, 1955), and the use of a 2-phase polymer system to extract bacterial spores from soil (Sacks & Alderton, 1961) are other examples of direct isolation of micro-organisms.

### 5.3 Indirect methods of isolation

### 5.3.1 Plating of untreated samples

(a) *Isolation from soil suspensions.* Micro-organisms occur in soil at differing concentrations. In order to isolate these from the soil and from one another, it is necessary to suspend the soil in a medium and/or diluting fluid, the degree of dilution being related to the initial concentration of organisms in the soil.

Generally, no attempt can be made to define the micro-habitats from which the organisms come, although in certain special cases it might be possible to prepare suspensions from readily identifiable and separable soil particles or roots. These isolation techniques also preclude the determination of the condition of the organisms being isolated, i.e., vegetative *v.* spore states. Isolation techniques giving information on this point are dealt with later (see p. 43).

(i) Soil plate method (Warcup, 1950). This technique allows isolation of fungi from a large number of separate soil samples without the labour of preparing and using the many dilution blanks required in the soil dilution plate method. Data on the relative frequency of occurrence of individual species can be obtained using this method.

Suitable small amounts of soil (see later) are taken from the main sample using a sterile needle with a flattened tip. Each small sub-sample is placed in a drop of sterile water in a Petri dish and dispersed. 8–10 ml of a molten agar medium at 45° are then added to each Petri dish, the dish being rotated gently to allow dispersion of the soil particles in the nutrient medium.

The dishes are incubated at an appropriate temperature (25° being commonly used) and observed daily for fungal growth. Attempts must be made to sub-culture and remove fast growing species as soon as they appear on the isolation plates, otherwise they will overgrow the whole plate.

In the early stages of an investigation of soil fungi, all colonies developing on the soil plates should be sub-cultured on to nutrient medium as soon as possible after their appearance on the plate. The replication of sub-culturing can be cut down only when the investigator can identify the most frequent forms on the initial isolation plates.

Care should be taken to ascertain the most suitable amount of soil to be placed in each Petri dish. Warcup used 0·005–0·015 g soil/plate, but if the soil contains only sparse fungal populations then larger amounts of soil should be plated. This can only be determined by preliminary experimental tests.

Warcup (1951) took a series of small samples from each horizon of each soil under study. Three soil plates were prepared from each sample. To obtain a positive presence record, a fungus had to appear on at least one of the three plates and consequently, profuse occurrence on all plates achieved the same score as a single presence on one plate. Other workers have prepared large numbers of plates (e.g., 50) from a bulk soil sample, and have recorded presence or absence of individual species on each plate.

The technique can be applied to a wide range of soils. It is quick and easy to use, and for these reasons it is useful in making initial surveys of soil mycofloras. However, the fungi isolated by this method tend to be those present as spores in the soil sample.

A special modification of this method for isolating nematode-trapping fungi has been described by Duddington (1954, 1955). Small amounts of soil or organic residues are scattered in a sterile dish and gently agitated with cooled, molten agar. The choice of medium is critical, for use of rich nutrients encourages luxuriant growth of other microbes which makes detection of trapping fungi difficult. A weak nutrient medium, such as weak maize meal extract agar or plain water agar should be used.

Plates are incubated (18–25°) for 2 weeks and then examined periodically with a microscope for the presence of nematodes and trapping structures. Observations may be carried out for a period of 1–2 months, as some species are slow to appear on plates. These fungi are most efficiently isolated into pure cultures by picking off spores from sporophores growing above the medium, using a fine needle and a binocular dissecting microscope.

(ii) Dilution Plate Methods. The soil plate method has been used almost exclusively for isolating soil fungi. Isolation of organisms with restricted colony growth (e.g., bacteria and actinomycetes) which do not spread out from soil crumbs, should be attempted by the use of soil dilution techniques. Often, isolations may be made most conveniently from dilution plates prepared when making biomass determinations. The method used to prepare the initial soil suspensions for these determinations is described on p. 65. If biomass determinations or counts of propagules are not required, the following modifications to the treatment of the soil suspension used in the dilution plate count may be made.

Soil/dispersing fluid ratio. It is not necessary to have an accurately known amount of soil in any given quantity of dispersing fluid, although there is probably no advantage to be gained from deviating markedly from the procedure outlined on p. 65.

Dilution of samples. A simple pour plate method can be used. A loopful of suspension is transferred to a vial containing 15 ml of suitable molten medium, held at 45°. The vial is rotated several times between the hands to disperse the inoculum. A loopful of this inoculated agar (or larger quantity if desired) is transferred to a second vial of molten agar and the procedure repeated. The number of vials used in the dilution series depends upon the concentration of propagules, but generally no more than three are needed. When all the vials have been inoculated, they are poured into Petri dishes and then treated as outlined in the description of the dilution plate count (p. 65).

The method is only slightly different from the soil plate method (p. 40) but is useful if a better dispersion of propagules, e.g., bacteria absorbed on soil particles, is required.

(iii) Surface plating. An alternative method for inoculating agar plates may be employed in which the soil suspension is spread evenly over the surface of medium previously poured into a Petri dish and allowed to set. This has the advantage that all the developing colonies are easy to pick off from the plates for subsequent purification; digging out colonies buried in the agar is eliminated. Also, the response of the entire population of isolates to different media can be assessed by use of the replica plating technique (Lederberg & Lederberg, 1952).

The chief disadvantage of surface plating is that colony growth may be less restricted on the agar surface so that colonies merge into one another in a short period of time. The relative importance of such advantages and disadvantages can only be determined by preliminary examination of the samples at hand.

A useful surface plating method is that described by James (1959):

Pour sufficient plates of the appropriate medium containing enough agar to give a hard medium on setting, e.g., if the normal setting concentration is 2%, use 2·5% agar.

Prepare a series of dilutions (see p. 65) but include in the diluent, half the concentration of agar recommended for solidifying a medium, i.e., if the normal setting concentration is 2%, use 1% agar.

Inoculate the surface of the solid agar with 1 ml of the diluted soil. Good distribution of colonies can be obtained by spinning the plate on a turntable as the inoculum is spread quickly with a sterile flattened loop or a piece of bent glass rod.

Incubate the plates in the most appropriate manner.

If trouble is experienced with excess inoculum not being absorbed by the agar, a smaller inoculum may be used. The size of inoculum needed will be partly dependent upon the dilution used, so preliminary tests on the range of dilutions required may be needed.

(iv) Isolation of antagonistic micro-organisms. Antagonistic microbes are present among random isolates obtained from almost any soil, using isolation procedures already described. However, their antagonistic properties are not usually discovered until after isolation into pure culture, when they are tested against other microbes. Various modifications to isolation procedures can be made that allow the detection of antagonistic microbes on plates, so that only suitable colonies need be transferred to pure culture. A variety of procedures have been described, but the basic steps involved are similar in all.

Selection of test micro-organisms. This depends upon the aims of the investigation. If a general survey of antagonism is required, test micro-organisms should include representatives of Gram-positive bacteria, Gram-negative bacteria, actinomycetes, fungi and yeasts.

Seeding of medium with test micro-organisms. The medium is seeded, whilst molten at 45°, with propagules of the test organism and mixed well. The medium should be chosen to allow good growth of both the test organism and soil isolates. It should be reasonably clear when solidified. Seeded medium in standard amounts is poured into Petri dishes and allowed to set.

Introduction of soil suspension. Suitably diluted soil suspensions (prepared as for the dilution plate technique—see Section 5.31b) may be introduced to the seeded medium in a number of ways. Samples may be pipetted into the seeded medium while it is still molten and mixed intimately with the test organism before pouring. This is suitable if the test organism is reasonably fast-growing. Alternatively, the test organism may be allowed to grow for a short period, e.g., 12 hours, before introduction of the soil suspension. In this case, the suspension must be applied to the surface of the solidified seeded medium. This can be done by pipetting it on to the surface followed by spreading with a bent glass rod, spraying on, or mixing it with a small volume of agar and pouring on as a layer on the basal seeded medium.

Detection of antagonistic microbes. After a suitable period of incubation, plates are examined for colonies which are surrounded by clear zones where the test organism has been inhibited or killed. Selected colonies can then be picked off into pure culture.

Examples of methods used to isolate particular groups of antagonistic micro-organisms are those described by Kelner (1948) for antibacterial isolates and Carter and Lockwood (1957) for isolates lysing fungal mycelium.

### 5.3.2 Plating of pretreated samples (growth precluded)

(a) *Washing techniques*

These techniques have been designed to achieve one or more of the following aims:

1. The separation from soil of its constituent particles (micro-habitats) before attempting to isolate micro-organisms.

2. The removal of microbial propagules which are loosely attached to these soil particles, leaving only those which are intimately associated with them. The removal of most spores of common soil fungi is achieved, increasing the chances of isolating forms present as mycelium, slower growing species and those inside organic particles. The application of these techniques to the isolation of bacteria and actinomycetes is more limited. Many soil bacteria are readily detached from their substrates and are best isolated from a soil water suspension (see p. 40). Actinomycetes may be isolated from washed material but it is difficult to obtain pure colonies from particles where other fast-growing, spreading organisms are present.

(i) Root washing technique (Harley & Waid, 1955). This method has been used to wash leaf litter as well as roots. After preliminary washing with non-sterile water, the material under study is placed into sterile screw-topped vials (*c.* 25 ml) containing approximately 10 ml of sterile water. The quantity of material placed in a vial should be such as to allow free movement of water in the subsequent agitation process. The vials are agitated in a mechanical, preferably horizontal, shaker. Hand shaking may be used if a mechanical shaker is not available. Agitation is normally carried out for 2-minute periods and after each the water is decanted off carefully to avoid loss of the material.

Decanted water can be collected in separate sterile containers for testing the efficiency of the method. For any study, it is essential to determine the minimum number of washings needed to remove most detachable propagules from the substrates under investigation. Material should be washed at least 30 times, the water from selected washings (e.g., 1–5, 10, 20, 30) being collected. The number of propagules in each of these washings can be determined by preparing a series of dilutions and using the dilution plate procedure outlined elsewhere (p. 64). If a graph of number of propagules removed against number of washings is plotted it will be clear how many washings are needed to reach a low, fairly constant rate of propagule removal. The efficiency of the method can be improved by one or more changes of the vials during the washing procedure, particularly if heavier soil particles remain in the vials after decanting.

Material washed for a suitable period is then transferred aseptically to a sterile container, e.g., a Petri dish containing filter paper. Sufficient drying of the material on sterile filter paper will discourage spread of bacteria in the incubation period. Material is then transferred aseptically to plates of a suitable solidified medium. The number and size of particles placed on each plate should be as small as possible for efficient isolation, but quantities used vary with the nature of the washed material and the aims of the investigations. Incubation and subsequent procedures are the same as those described for the soil plate method (p. 39).

This method is most suitable for the isolation of fungi growing on the larger solids in soil, e.g., roots, leaf remains. In such cases it provides useful information and requires no elaborate, specialized apparatus. For smaller soil particles or those with a low specific gravity, its value is more limited as it is difficult to achieve fast, efficient removal of washing water without loss of material.

(ii) Soil washing technique (Parkinson & Williams, 1961; Williams, Parkinson & Burges, 1965). This method and modifications of it have been used to wash soil. Unlike the method of Harley & Waid (1955), no attempt is made to separate out particular types of substrate material before washing and whole soil samples are washed.

Samples are placed in alcohol sterilized perspex boxes which contain a number of sieves of graded size, chosen according to the material studied. Soil is placed on the upper, largest sized, sieve. Sterile water is added and the soil subjected to agitation by passage of sterile compressed air. After 2-minute periods of agitation, the water is run off and fresh water added. Thus soil can be given a number of serial washings.

The washing water can be collected if needed. For testing the number of washings needed for a soil (frequently 20–40) selected washings should be collected and tested by the same procedure used for the root washing technique. If the nature and quantity of propagules removed by washing are of interest, all washings can be collected and bulked before dilution and plating.

On completion of the requisite number of washings, discrete soil particles are found, graded according to size on the various sieves. These can be removed aseptically by rinsing each sieve separately in a Petri dish containing a few ml of sterile water. Particles can then be picked out, dried and plated, following the procedures outlined for the root washing technique.

If more than one soil sample is to be studied, it is possible to build a battery of boxes each receiving air and water from a common supply (Williams *et al.*, 1965). Other modifications of the apparatus have been described. An apparatus having a continuous flow of water, with the agitation being provided by either hand or mechanical shaking has been described by Gams & Domsch (1967), whilst a fully automated washing device has been used by Hering (1966).

The soil washing technique achieves two main objectives. First, different types of soil particles, including smaller, lighter ones not suited to study by the Harley & Waid technique, e.g., different sized mineral grains, root fragments, and animal remains, are separated and can be plated as discrete units. Secondly, as with the root washing technique, fungi intimately associated with particular substrates can be isolated. If both washed particles and washing water are eventually plated, a more complete picture of the fungus population of the soil is obtained, than when the soil plate method (p. 39) or the dilution plate method (p. 40) are used. It is possible to wash a number of soil samples simultaneously, in different boxes, each sample receiving the same standardized treatment.

Difficulties are encountered when soils containing much colloidal material (clay or humus) are washed. In such cases, removal of propagules by the washing process is less efficient, and even if 40 washings are given, a low constant rate of propagule removal may not be achieved.

In deciding whether or not to build a washing apparatus, the aim and scope of the work should be considered. If an extensive study of soil fungi including regular sampling is envisaged, the effort involved in building such an apparatus would be worthwhile. On the other hand, in a relatively short term study, it may be sufficient to use the Harley & Waid technique for soil washing, despite its limitations for dealing with smaller, lighter materials.

(b) *Surface sterilization*

The application of washing techniques to the isolation of micro-organisms from a range of organic material has been described. However, it is not always possible to ascertain the origin of isolates obtained by this method (i.e., whether they originate from hyphae closely adhering to the surface or from within the tissues of the inoculum).

The careful application of microbiocidal agents to the surface of organic material taken from soil can allow isolation of the micro-organisms colonizing internal tissues, all surface-living forms being killed. This technique has been commonly used by plant pathologists to study seeds, leaves, roots, root nodules, mycorrhizas, and stems, even though it is sometimes difficult to ensure complete removal of surface propagules.

A variety of surface sterilizing agents have been used, including mercuric chloride solution with or without added silver nitrate, calcium hypochlorite solution, and equal parts of absolute alcohol and 20 vol. hydrogen peroxide.

Structures from which micro-organisms are to be isolated are placed, for a standard time, in a surface sterilant of standard concentration. Upon removal from the solution, all residual traces of the sterilant must be removed by serial washing with sterile water. Then pieces of the surface sterilized material are blotted dry and plated onto nutrient medium. Micro-organisms growing from the plated pieces of material should be sub-cultured onto nutrient medium slopes and identified.

Each phase of this generalized procedure must be carefully tested to ascertain the appropriate concentration of sterilant, time of application, and requirements for complete removal of sterilant. These will vary for different materials.

Examples of times of sterilization and concentration of sterilant are as follows:

1. Sodium hypochlorite solution with 1·6% available chlorine for times varying between 1 to 20 minutes.

2. Mercuric chloride (0·1%) for times varying between 1 and 5 minutes.

3. Nance's solution (1 g mercuric chloride, 5 ml 'Teepol', 1 litre distilled water) for times up to 10 minutes.

To test for removal of a sterilant, place pieces of surface sterilized and washed material on to agar plates which have been seeded with a spore suspension of a test fungus. If regions of inhibition of fungal growth occur around the previously surface sterilized material this shows that surface sterilizing fluid is still present on the material.

Particular caution must be taken in the application of surface sterilants to delicate plant structures (i.e., fine roots). Spongy and pitted materials are also difficult to deal with because of the difficulty of removal of residual surface sterilant.

When surface sterilizing delicate tissues, it is frequently necessary to assess the depth of the cell killing action of the sterilant. This can be traced with a vital stain (e.g., tetrazolium salts): A neotetrazolium solution (1 mg/ml) and a phosphate buffer solution (pH 6·0) are made up. One part of each of these solutions is added to 1 part of water and aliquots of the resulting mixture are placed in Thunberg tubes containing the material to be tested. These are evacuated and kept at 33°C overnight. The roots are examined under the microscope when living cells should show a deposit of red pigment.

In view of the general toxicity of many surface sterilizing agents the need for caution in the use of such materials must be emphasized.

While this method has been applied mainly to the study of fungi within plant tissues, it can be used to examine bacteria which occur in almost pure cultures within plant structures, e.g., root nodules.

## (c) *Fragmentation techniques*

These techniques have been used for the following reasons:

1. the isolation of micro-organisms from the inner regions of organic substrates, e.g., peat, when they are often combined with surface sterilization techniques.

2. the dispersion of materials on which there is concentrated microbial growth, facilitating the isolation of a wider range of forms when the material is plated.

3. the detection of actinomycete mycelium in soil.

(i) Dispersion of material and isolation from inner regions of tissues. This involves the physical fragmentation of organic matter. The material must first be separated from the soil; how this is done depends on the nature of the material. However, it is essential to wash the material, at least crudely, to remove extraneous soil particles before fragmentation. A large root can be rinsed under a tap and then given a few washings with sterile water (Harley & Waid, 1955), while small particles may be dealt with by a soil washing technique (Parkinson & Williams, 1961). Having separated and washed the organic material, it may be fragmented, macerated or dissected.

Root fragmentation (Warcup, 1959; Clarke & Parkinson, 1960). Separation of organic tissues into relatively large pieces (c. 1 mm) can be achieved by placing washed material in a Petri dish containing a single drop of sterile water. Pieces are cut up with a sterilized scalpel and finally dissected into as many small fragments as possible using two sterile needles. A suitable molten medium is poured into the dish and the fragments dispersed in it. This method requires no elaborate apparatus and is simple to perform. It is suitable for dealing with soft tissues which are easily fragmented but not for hard tissues. Some dispersion of the material is achieved, increasing the chances of isolating slower growing forms in the presence of their faster growing competitors, e.g., bacteria *v.* actinomycetes and fast *v.* slow-growing fungal genera. It is not usually possible to determine the original location of the isolated organisms.

Maceration (Stover & Waite, 1953; Clarke & Parkinson, 1960). A dispersion of organic material can be achieved by grinding it in a mortar, macerating it with a mechanical blender or both. The choice of method is governed by the type of material. Clarke (1961) described a technique for macerating the soft roots of *Allium*. After suitable separation and washing procedures, 1 mm lengths of root are transferred to 5 ml sterile water in a sterile $2\frac{1}{2}$in. mortar and covered by a Petri dish lid to prevent contamination. The pieces are crushed with a sterile pestle for 2–3 minutes, transferred to a sterile blender bottle and blended for 10 minutes at high speed with an M.S.E. Atomix blender. Aliquots of the resulting suspension are then transferred to a plate and incorporated into a suitable medium. The aims of such methods are the same as those of the root fragmentation techniques. However, a higher degree of dispersion is obtained and they are applicable to a wider range of materials. Although the exact origin of colonies developing cannot be determined, propagules of micro-organisms present inside organic matter are released and can be isolated along with those originating from surfaces. It is possible to combine such fragmentation techniques with surface sterilization procedures and thereby get a more accurate picture of the internal colonizers. Two possibilities must be borne in mind when using a fragmentation or maceration technique. Any assessment of number or frequency of organisms isolated is likely to be inaccurate as the fragmentation process can increase or decrease the number of viable units of fungal and actinomycete mycelium and bacterial cells. Also, Clarke & Parkinson (1960) have shown that maceration may release substances inhibitory to the growth of micro-organisms; consequently, macerates of sterilized organic matter should be tested against representatives of the microbial groups under study.

Root Dissection (Waid, 1957). The methods described so far do not usually give any direct information on the points of origin of isolated micro-organisms within tissues. With some of the larger and more defined pieces of organic matter in soil, e.g., certain plant roots, it is possible to break them up more carefully and plate out recognizable parts of their structure. A technique for dissection of rye grass roots has been described by Waid (1957). Roots previously washed by the method of Harley & Waid (1955) are placed in dishes of sterile water and dissected with a sterile scalpel and forceps under a binocular microscope. The outer cortex and stele cylinder fragments are separated and plated out, either immediately or after further washings. The feasibility of applying dissection techniques largely depends on the nature of the material to be studied and the apparatus available. If a good binocular microscope together with some form of micro-manipulator are available, many materials could be treated this way.

(ii) Detection of actinomycete mycelium in soil. A method for detection of actinomycete mycelium in soil which relies on the differential response of spores and mycelium to the abrasive action of coarse soil particles has been described by Skinner (1951). The method described here is a modification of the original.

About 2 g of soil are placed in a sterile blender bottle (25 ml) and 10 ml sterile water added. The mixture is blended vigorously for 60–90 minutes. At 5-minute intervals, 1 ml samples are removed from the suspension, diluted (2 or 3 10-fold dilutions are usually sufficient) and incorporated into a suitable medium for selective growth and counting of actinomycetes. A comparison of numbers of actinomycetes in the 1 ml aliquots after different periods of blending is then made and the ratio of the *highest count obtained* : *last count obtained* calculated. A ratio of 1 : 1 or less indicates that the vast majority of propagules are spores which are not killed by the prolonged abrasion with soil particles during blending. A ratio of greater than 1 : 1 indicates the presence of live mycelium in the soil sample and the greater the ratio, the higher is the proportion of mycelium in the soil. Live mycelium is eventually killed by blending and abrasion, and hence counts obtained in later samplings are progressively reduced. Figures obtained cannot be related to biomass in soil.

With natural, unamended soils, where a large proportion of actinomycete propagules are spores, it is unusual to obtain a ratio greater than 2 : 1. If the soil under study contains a low proportion of sand grains, it may be necessary to add sterile sand to the sample before blending. In comparative studies, attempts should be made to use samples with approximately the same proportions of sand particles in them.

(d) *Heating and desiccation of samples*

(i) Pasteurization of soil samples. If soil suspensions are heated before attempts to isolate organisms are made, then only the heat resistant forms will survive. This is useful when one wishes to isolate micro-organisms which are markedly different in their heat resistant properties from all other groups. The isolation of *Bacillus* and *Clostridium* is possible, bearing in mind that the colonies that develop during isolation will have arisen from spores, rather than from vegetative cells which are not so heat resistant.

A useful method is as follows:

Prepare a soil dilution series as described on p. 65.

Place the tubes containing the diluted soil suspension (or the original soil suspension) in a water bath at 80°C. After allowing 1 minute for the temperature of the suspension to reach 80°C, leave the tube in the bath for a further 5 to 10 minutes, perferably applying intermittent gentle shaking. For isolation of clostridia, heating at 80° for 15 minutes has been suggested.

Remove the tubes and plate out the suspension on a suitable medium, e.g., normal isolation medium, nutrient agar, Mishustin medium, reinforced clostridial medium, and incubate the cultures at 25° until colonies are evident.

If desired, this procedure may be varied by testing the effects of different temperatures and different times of exposure to heat. This will depend on the nature of the soil population, as spores of different species have different heat resistance properties.

(ii) Heat treatment for isolation of ascomycetes. Warcup & Baker (1963) have described a method for pretreating soil which results in the death of most fungi and the germination of dormant ascomycete spores which then grow on isolation media.

2·5 g soil is immersed in 60% alcohol for 6–8 minutes. This kills off many other fungi, but may be omitted if thought unnecessary. The resulting suspension is added to a water blank to give a 1 : 100 dilution and heated at 60°C for 30 minutes in a water bath. One ml samples are added to Czapek Dox Agar + anti-bacterial antibiotics, poured into Petri dishes and incubated.

Limitations of the techniques include the variation in reaction to high temperatures noted above, and the impossibility of knowing whether vegetative cells inside organic matter particles are affected in the same way as those in suspension. The techniques may be used to count spores in the soil, using the dilution plate count procedure.

(iii) Drying of soil samples. Drying soil samples before plating or dilution has been recommended as a method for killing bacterial populations when they might interfere with the isolation of actinomycetes and fungi. However, drying has severe effects on some fungal mycelium (McLennan, 1928) and spores (Warcup, 1960) and may also affect actinomycetes. Drying can only be recommended if no other way of suppressing unwanted micro-organisms is available. A convenient method is as follows.

Place soil in a Petri dish, breaking up any large clumps, and put the dish into a vacuum desiccator containing calcium chloride. Evacuate the desiccator and leave the soil samples to dry for three days. Plate out or dilute the dried soils in an appropriate way.

A simple method for sandy, more easily dried soils would be to place them at 45° in an incubator for times varying from 4–24 hours.

### 5.3.3 Plating of pre-treated samples (growth allowed)

(a) *Pre-treatment by chemical methods.* Frequency of occurrence of micro-organisms in soil is only a crude guide to their importance in soil processes since many important reactions are carried out by organisms present only in small numbers. Isolation of these organisms presents special problems as samples have to be pretreated or 'enriched' so that the organisms to be isolated grow at the expense of all other types. These enrichment techniques are excellent for the isolation of organisms belonging to particular physiological groups. However, they have some general limitations since they provide no information on the relative abundance of organisms in natural soil and do not always allow isolation of all the organisms participating in a particular process. More frequently, a narrow spectrum of forms emerges, capable of adaptation to the changed soil conditions. Addition of particular substrates in different physical forms is recommended so that a broader picture is obtained. Three main methods of adding substrates will be considered in this context, (i) addition of powders and liquids to soil under static conditions, (ii) addition of localized substrates to soil under static conditions, and (iii) addition of substrates to soil by percolation.

(i) Addition of powders and liquids to soil under static conditions. Powdered substrates and liquids may be blended intimately with soil and the resulting mixture incubated under the desired conditions. Isolations from this mixture may be made by many of the techniques already referred to, e.g., soil crumb, soil suspension methods, etc. These techniques have been most widely applied to the bacteria, but there is no reason why many of them should not be used to enrich soils for peculiar groups of fungi and actinomycetes present in small amounts in the soil.

Winogradsky soil block technique (Rubenchik, 1963). A layer of washed charcoal is placed at the bottom of a Petri dish and covered with the soil to be investigated. This soil should have been sieved and mixed with 5% starch and water to obtain a thick paste. An alternative, and perhaps better procedure (Mishustin, 1954) is to mix the soil with a solution containing 0·1% $K_2HPO_4$ and 5% mannitol. The soil surface is then smoothed with a wet glass slide and the dish incubated at 30° for 48 hours. It is advisable to place a small glass tube in the charcoal bed so that air may pass into the system. Colonies develop on the surface of the soil and may be picked off and cultivated on a nitrogen-free medium (see p. 108). The basis of the method is to supply organic material in the form of a carbon source and to omit nitrogen. Growth of organisms requiring soil nitrogen will therefore be low, whilst growth of bacteria such as *Azotobacter* which can utilize atmospheric nitrogen will be high.

(ii) Addition of localized substrates to soil. It is often useful to add substrates to soil in the form of clearly localized pieces, so that isolations are made only from the substrate and not from the whole soil mixture and thus the chances of successful enrichment will be increased. In some instances direct observations can be made of the colonization process immediately prior to isolation, making this form of enrichment doubly useful. These techniques have been used most frequently for isolating soil fungi. The reasons for this are not clear, but it is reasonable to suppose that in many soils, organisms which can grow through soil to the substrate, e.g., fungi, stand a better chance of becoming established than organisms growing as restricted colonies, e.g., bacteria, which would have to be present very close to the substrate before enrichment could occur.

The buried strip technique. Several versions of this technique have been described. They usually involve the burial of a transparent form of the substrate. During colonization, pieces of the substrate are recovered from soil and observed under the microscope to establish the broad pattern of colonization, whilst other pieces are plated out on agar medium with or without washing treatments. Micro-organisms which grow from the strips are then isolated into pure culture. Substrates examined in this way include cellulose, chitin, cutin and lettuce leaves, i.e., both pure and mixed materials. Tribe's original technique (Tribe, 1957) for burying cellulose is described in detail, with some amendments. Details of procedures involved in dealing with other substrates can be found in papers by Tribe, 1963 (lettuce leaves), Okafor (1966) and Baxby & Gray, 1968 (chitin) and Gray & Lowe, 1967 (cutin). Cellulose may be obtained in the form of sheet cellophane (British Cellophane Co., type P.T. 300) and cut into pieces about $1\cdot0 \times 0\cdot5$ cm. These pieces should be washed in at least two changes of distilled water to dissolve out any added plasticizers which might affect microbial growth. The strips are placed on 22 mm square cover slips and the excess water drained off. If desired, the strips may be dried on microscope slides. The cover slips are buried in soil in pots or plastic boxes kept in the laboratory. These pots may be kept under desired moisture and temperature regimes, though 60% of the moisture holding contents of the soil and 25°C are frequently used for general isolation programmes. Strips may be removed at different time intervals after burial so that enough slides should be buried to provide sufficient replicates during the whole experimental period. Strips may be observed with or without staining. Picronigrosin in lactophenol (Smith, 1956) is a useful stain. Isolations are made from unstained strips. These may be washed before plating using the Harley & Waid technique (see p. 43) if desired. After washing, the strips are cut into 2 mm squares and plated on a suitable medium. If bacteria and actinomycetes are to be isolated, it may be necessary to macerate the strips and plate out dilutions of the macerate on media containing antifungal antibiotics (see p. 106) for, if the strips are plated directly, fungi may continue to grow on the strip which contains no antifungal antibiotic and prevent the spread of bacteria and actinomycetes on to the media. Another useful isolation procedure is to bury a triple 'sandwich' of cellophane in the soil. Fungi isolated from the centre of the sandwich may be presumed to be cellulolytic since they have grown through the outer layer first; they are often present in discrete patches which may be isolated more easily.

The buried pellet technique. Many substrates cannot be obtained as transparent strips because they are soluble, opaque or granular. The colonization of these substances can be followed by incorporating them into pellets which are then buried in soil. These are useful for studying not only the localized development of microbial populations, but also the production of metabolites in soil. Because they are opaque it is difficult to observe organisms on their surfaces, although staining with fluorescent dyes and observation under incident ultra-violet or scanning electron microscopy may help (see pp. 25–26). A convenient technique is described by Webley & Duff (1962). Soluble or granular materials are ground and mixed in the required proportion (usually 1 : 1) with a suitable insoluble base, e.g., kaolinite. The material is placed in a steel mould and subjected to pressure. Usually $0\cdot2$ g of mixture subjected to a load of 5000 lb, is used to produce a disk-shaped pellet about $9\cdot5$ mm

E

diameter and 1·35 mm thick. The nutrient pellets are buried in soil together with control pellets containing no nutrients. After incubation for the required period of time the pellets are removed, shaken for 5 minutes in a known volume of sterile water and dilutions made from the resulting suspension.

Baiting techniques. Another similar form of enrichment is baiting. This is most frequently used to isolate special groups of micro-organisms, e.g., aquatic phycomycete fungi from water or soil suspensions. Sparrow (1957) has described a useful technique which could be modified for different purposes. Soil samples are collected in sterile boxes or tubes. After return to the laboratories, a generous tablespoonful of soil is placed in a deep sterile Petri dish containing enough sterile water to allow a free water surface to form above the soil. Each plate made in this way may be baited with different materials, e.g., cellophane, boiled grass, pine pollen, boiled hempseed, shrimp chitin, insect wings, defatted hair or snake skin. Cultures should be incubated for at least a month, and examined frequently as different waves of colonizers appear. Fungi may be obtained in pure culture, using conventional methods of purification. Willoughby (1968) has also used pollen grains as bait for aquatic actinomycetes. Another useful baiting technique for nematode-trapping fungi has been described by Cooke (1962) and has been called the agar disc method. Discs (1 cm diameter and about 3 mm deep) are cut out of plates of weak maize-meal-extract agar. These are placed on slides in a row (4 per slide). Each slide is placed in a 15 cm Petri dish of soil, so that the discs are about 1 cm below the soil surface. At weekly intervals, slides are removed and replaced with others holding fresh discs. The discs extracted from the soil are washed, while still on the slides, with a fine jet of water and examined microscopically, unfixed and unstained. During decomposition of organic matter, nematodes and trapping fungi appear on the discs. A comparative assessment of activity can be obtained by recording the frequency of occurrence of trapping fungi on the discs, e.g., as number of discs colonized, number of traps found per disc.

(iii) Addition of substrates to soil by percolation. In the enrichment technique described so far, substrates are added to soil at the start of an experiment and organisms isolated after a series of changes have taken place. However, soil can be percolated continuously with a nutrient solution and organisms isolated from the solution that has passed through the soil. The main advantage of this method over the static methods is that the physical condition of the soil, e.g., aeration and water content, remain relatively constant, albeit at somewhat unusual levels, and that samples can be taken from the perfusate without disturbing the soil. Experiments on the rates of processes can be carried out in addition to isolation of micro-organisms.

Cripps & Norris (1969) have described a simple form of perfusion apparatus which is ideal for isolating organisms capable of utilizing different substrates. The apparatus is made from two circular plates (15 cm diameter) connected together by five 50 cm metal rods. This cylinder may be rotated about its long axis at 1 rpm by an electric motor (Parvalux Electric Motors Ltd., Bournemouth, England). Soil is placed in a glass tube (6 × 1 cm), plugged at both ends with glass wool; 0·5 g soil is usually sufficient. The ends of the tube are joined by polyvinyl chloride tubing in which is placed 8–10 ml of the perfusing medium. The composition of this medium will depend upon the organisms that are to be isolated. The completed loops are fitted around the cylinder which will accommodate about 30 loops and the apparatus switched on. After the soil has been perfused for several days or weeks, the liquid in the tubes may be sampled with a syringe or by breaking open the tubes and samples plated out on a suitable medium. A more complex version of this apparatus suitable for metabolic studies is described on p. 89. Although this method seems attractive because of its simplicity, it does not incorporate certain important features possessed in the more complex methods. For instance, there is no constantly renewed air supply and it is probably only marginally superior to the straight forward enrichment of a medium in a shake flask unless it is desired to grow organisms which thrive under reduced oxygen tensions.

(b) *Pretreatment of soil by physical methods.* Organisms vary in their response to changes in their physical environment and it is possible to pretreat soil samples to allow the isolation of those forms capable of growth under different conditions. Soil may be kept at different moisture tensions or under different gas mixtures, but it should be remembered that resting propagules resistant to change in the environment may survive this treatment. This being so, it is essential, subsequently, to isolate on media in which the conditions of pretreatment are maintained, or to isolate using methods selective for vegetative phases of growth, e.g., direct hyphal isolation, soil washing methods. In practice methods involving physical pretreatment are rarely used since good results can be obtained by incubating the isolating medium under defined conditions.

(i) Constant moisture regime. Soils can be maintained at more or less constant moisture tensions by incubating them in controlled humidity cabinets. This method is most suited to maintenance of soils at low moisture contents. Soil is allowed to equilibrate in a humidity cabinet (e.g., apparatus made by Townson & Mercer Ltd., Croydon, England) in which forced circulation of air takes place. The requisite humidity is obtained by equilibrating the soils against different salt water solutions, a process which takes at least two weeks. The salt solutions are placed in a tray immediately above an air fan. A suitable range of salts is $CaSO_4.5H_2O$ (98% R.H.), $Na_2HPO_4.12H_2O$ (95% R.H.), $KNO_3$ (94% R.H.), $K_2HPO_4$ (92% R.H.), $NH_4Cl$ (79% R.H.) and $CaNO_3.4H_2O$ (52% R.H.) (see also Scott, 1956; Dubey, 1957; Durbin, 1965).

(ii) Temperature. Soils may be pretreated by incubation at different temperatures. Problems may be encountered as the soils dry out during incubation especially when high temperatures are used. In these cases, it may be necessary to incubate them under constant humidity conditions as well.

(iii) Gas mixtures. The most familiar pretreatment of this type is the incubation of soil under anaerobic conditions, usually in one of the commercially available anaerobe jars. However, it is possible to incubate soil in a variety of gas mixtures containing varying percentages of oxygen, carbon dioxide and nitrogen. These may also be obtained commercially.

# 6
# Biomass Measurements

## 6.1 Introduction

One of the features common to most soil microbiological studies is the counting of cells or the measurement of hyphal lengths present in a soil sample. Such counts and measures are intended to provide evidence on the relative importance of organisms in the soil, but they are of limited value in this respect, and must be converted to weights of living material (biomass) to allow true comparison. Thus, it is possible to compare usefully the weight of bacteria and fungi in a sample, but not the number of cells and hyphal lengths. However, biomass determinations are of limited value since in most cases they refer to standing crops of organisms and not to cumulative totals or rates of production over a period of time. Nevertheless, they are often used to provide a rough guide to changes in populations in a soil profile or over a period of time.

Where biomass is calculated from cell counts a further problem is encountered in obtaining an acceptable figure. Direct counts are sometimes a thousand times larger than viable counts, and although many reasons can be put forward to account for this, they do not help to decide which count is the most accurate. The most accurate count would be a direct microscopic count in which living cells could be distinguished from dead ones but no really reliable method for doing this exists and even if it did, it would still be necessary to find some way of distinguishing bacterial cells from actinomycete spores and fragments of organic matter. Methods for determining biomass that are not based on counting propagules are therefore desirable, and one of these, based on the utilization of freshly killed cells by living micro-organisms, is given later.

## 6.2 Expression of results

Counts and biomass determinations are usually made with reference to standard weights of soil. Thus, it is common to see both expressed as numbers or weights per gram of soil, most frequently, per gram of oven dry soil. Whilst this probably remains the most convenient method of expressing results, other methods may add considerably to the value of the data.

### 6.2.1   Weight of a soil constituent
If it is suspected that the efficiency or method of utilization of some soil constituent by the microflora varies within a set of samples, it is useful to express the determinations in terms of the weight of this constituent, e.g., organic matter, carbon, nitrogen, phosphate, etc.

### 6.2.2   Volume of soil
Some soils differ substantially in specific gravity, e.g., organic and mineral soils, and in such cases it is advisable to express results per cm$^3$ of soil as well as per gram. This gives a better idea of the degree of crowding or separation of microbes within the soil.

### 6.2.3   Specific surface area of soil or its constituents
A more accurate idea of the degree of crowding may be obtained by expressing results in terms of the specific surface area of soil available to the micro-organisms, i.e., the summation of the surface area of all the individual soil particles in the sample. This is particularly useful when comparing habitats such as root surface and soil where a comparison on the basis of weight or volume is meaningless. Surface area of large particles may be determined by direct measurement, using the standard formulae.

Surface area of a sphere $\quad\quad\quad = 4\pi r^2$

Surface area of an oblate spheroid $= 2\pi a^2 + \dfrac{\pi b^2}{x} \log_e \dfrac{1+x}{1-x}$

$$x = \text{eccentricity} \left( x^2 = 1 - \frac{b^2}{a^2} \right)$$

$$a = \text{length of major axis}$$

$$b = \text{length of minor axis.}$$

### 6.2.4   Surface area of ground
In studies of energy flow in the soil ecosystem, energy input is often expressed in terms of Kcals per hectare of ground, because the main energy input is sunlight. Consequently, it may be necessary to calculate the numbers of soil organisms present with reference to the ground surface area, and to do this the

depth of soil and variations in biomass with depth may have to be determined. This is particularly necessary since soil zoologists routinely express their biomass data in this way.

### 6.2.5  Choice of method
The method chosen will depend upon the purpose of the study, but it is important to give sufficient data concerning the soil to allow other workers to express the results differently should the need arise.

### 6.3  Conversion of measurements to biomass

By use of the appropriate techniques, the number of bacterial cells, actino-mycete and fungal spores and total lengths of fungus and actinomycete mycelium in a soil sample can be found. This information can then be used to estimate the biomass of microbes in a soil. There are a number of ways in which this can be done.

### 6.3.1  Conversion of cell and spore counts
To convert these counts to biomass, the weight of an individual cell or spore must be calculated. This can be estimated as dry weight or fresh weight.
(a) *Dry weight*. A culture of a soil isolate is prepared in a suitable medium. For bacteria, liquid media are used, while solidified media are most suitable for producing fungal and actinomycete spores. Once maximum growth is obtained, the bacterial cells are harvested by serial washing and centrifuging in sterile distilled water, to remove all traces of the medium. Fungal and actinomycete spores can be scraped off the surface of the medium into distilled water and passed through a 'filtration system to remove pieces of mycelium before washing and centrifugation.

On completion of washing the cells are resuspended in a known volume of distilled water to give an homogeneous suspension. Aliquots are removed from the suspension and suitable dilutions of them placed in a Helber cell, enabling the total number of cells in the suspension to be calculated. The dry weight of cells can be found by passing the suspension through a sintered filter pad (dry weight previously found) and drying to constant weight at 60°C. Alternatively the water may be removed by evaporation. This may be achieved by heating the suspension in a watch glass or evaporating basin at 105°C. If freeze-drying apparatus is available a more controlled evaporation can be obtained under vacuum with a desiccant such as $P_2O_5$.

(b) *Fresh weight*. Fresh weights of cells can be calculated by measuring their density and volume. A crude estimate of the buoyant density of microbial cells can be determined using the following procedures. Dense cell or spore suspensions are prepared as described for dry weight determinations. Small volumes of suspension are mixed into solutions of a salt (e.g., CsCl) of known density (ranging from about 1·0 to 1·8) and centrifuged. In solutions with a density lower than their own, the cells are sedimented, while in solutions with a higher density they rise to the surface. In solutions with a density close to their own, the cells remain evenly distributed throughout the solution. One disadvantage of this method is that cells often form clumps which react differently from individual cells.

The volume of cells can be calculated from determinations of their mean dimensions, bearing in mind their approximate shape, e.g., spherical, oblong, cylindrical. The mass of one cell can be calculated using the formula:

$$\text{Mass} = \text{Density} \times \text{Volume}$$

Thus for a spherical bacterial cell of 1·0 $\mu$m diameter and a density of 1·5 the mass is calculated as follows:

$$\text{Cell volume} = \frac{4}{3}\,\pi(0·5)^3 \times 10^{-12}\ \text{cm}^3$$

$$\text{Mass} = \frac{4}{3}\,\pi(0·5)^3 \times 10^{-12} \times 1·5\ \text{g}$$

$$= 0·77 \times 10^{-12}\ \text{g (approx. calc.)}$$

Similar estimates are sometimes made by presuming that the average bacterial cell has a density of 1·5 and a volume of 1 $\mu$m$^3$. Mass then equals $1·5 \times 10^{-12}$ g, which is quoted by Alexander (1961) as the weight of moist viable tissue in an average bacterial cell.

### 6.3.2  Conversion of mycelial lengths

(a) *Dry weight*. Fungus or actinomycete mycelium which is non-sporing can be produced in liquid medium which is shaken during the incubation period. This is harvested and washed as described for cells and spores. Dry weight determinations can also be carried out using the methods previously described. However, it is necessary to find the total length of mycelium which is weighed and this is best done by preparing Jones & Mollison slides (see section 6.41) from aliquots of the suspension before drying. It is thus possible to calculate the dry weight per unit length of mycelium for a fungus or actinomycete isolated from soil.

(b) *Fresh weight*. Measurements of the density of mycelium are difficult as clumping is very marked. However, fresh weights can be calculated on a theoretical basis. It can be assumed that mycelium is a perfect cylinder (volume $= \pi r^2 l$) and that its average density is similar to other microbial cells (1·4–1·5). Its volume can be calculated from the mean of measurements of its diameter, and its length. Mass is then estimated using the average values of density and volume.

$$\text{e.g., Total length of mycelium} = 60 \text{ m/g soil}$$

$$\text{Mean diameter} = 4\cdot0 \ \mu m$$

$$\text{Assumed density} = 1\cdot5$$

$$\text{Vol.} = \pi r^2 l$$

$$\text{Mass} = \pi r^2 l \times 1\cdot5 \text{ g}$$

$$= 1\cdot13 \times 10^{-3} \text{ g}$$

### 6.3.3  Problems in application of techniques for soil biomass determinations

All methods for converting counts and measurements of soil microbes to biomass are carried out on organisms grown in artificial media. Hence inaccuracies will arise if the microbe's dimensions and mass are not the same as when it grows in soil.

Also measurements of organisms in soil are obtained by direct observation methods which do not allow identification. It must be assumed that the microbes, grown in culture and on which biomass determinations are based, are representative of those observed in the soil.

Finally, if total microbial biomass is required, it is not feasible to obtain conversions for all microbes in a soil, even if they have been isolated in pure culture. Once again it is necessary to extrapolate information obtained on a few microbes to the whole population.

## 6.4   Measuring and counting

### 6.4.1   Agar film technique (Jones & Mollison, 1948)

This method allows the assaying of lengths of fungal mycelium and/or numbers of bacterial cells in unit weights or volumes of soil. Comparative studies on assessment of fungal biomass in soil samples have shown this method to be the least prone to experimental error (Nicholas & Parkinson, 1967).

In using this technique the amount of soil normally chosen is between 0·5–4·0 g. The exact amount, which varies from soil to soil and from horizon to horizon, must be determined by preliminary investigation.

The soil sample is ground using a mortar and pestle and using 4–5 changes of sterile water (5 ml/change). The time for the first grinding is 5 minutes and for each subsequent grinding 2 minutes. Thus 20–25 ml of a soil-water suspension are obtained. In the preparation of this suspension all the soil particles are poured into a collecting container. Molten 1·5% agar (Oxoid No. 3) is added to the soil-water suspension to give a known dilution of the original soil sample (1 g soil in 50 ml). The dilution used again depends on the mycelial content of the soil under investigation and must be determined by preliminary investigation.

The soil-water-agar suspension is kept at 50–60° prior to preparation of agar films, agitated and then allowed to settle for 5–10 seconds. A small amount is pipetted from approximately 1 cm below the surface of the suspension and placed on the 0·1 mm deep platform of a haemocytometer slide. The platform is covered with a cover slip, taking care that the suspension does not overflow from the haemocytometer platform, and a 5 g weight put on the coverslip. The agar film is allowed to solidify.

After this, the slide is immersed in distilled water, the cover slip removed, the surplus agar in the moats of the slide cut away with a razor blade and the agar film floated off on to an ordinary microscope slide.

The preparation is allowed to dry at room temperature, and is stained with phenolic aniline blue for 1 hour (see p. 21), washed (3–4 changes of 98% alcohol) and mounted in Euparal. For some soils, observation of unstained films using phase-contrast microscopy is the best method.

The number of fields per slide and the number of slides observed must be chosen to allow proper statistical treatment of the data obtained.

Measurement of hyphal lengths/field is made either by projecting the images of the fields examined onto paper and using a map measuring device, or by using a camera lucida apparatus and measuring the drawn fragments.

Thus lengths of mycelium/unit weight or volume of soil can be assessed by taking into account the area of fields observed, the depth of agar film, the number of fields viewed, the total volume of suspension observed and the degree of dilution of original soil sample.

This method does not allow distinction between live and dead fungal hyphae. Some authors have claimed that stained hyphae (using phenolic aniline blue as the stain) are living hyphae, but this proposition has not and cannot be adequately tested. Another disadvantage of the method is that it does not give any indication of the relation between soil fungi and their micro-habitats since the basis of the technique is a disruption of the soil fabric. It is not usually applied to soils of high organic matter content in which it is difficult to release microbial cells from the organic fragments.

It does allow accurate, absolute, quantitative assessments of amounts of total living plus dead mycelium and bacterial cells in soil samples. Despite its disadvantages, it is one of the better methods for assessing lengths of fungus mycelium in soil samples.

### 6.4.2 Stained smears

(a) *Staining procedure*. Soil may be ground in a pestle and mortar (or blending machine) with water, a drop of the suspension spread over a 1 cm$^2$ area on a clean grease-free microscope slide and allowed to dry. After brief heat fixation, the slide may be stained with a variety of stains (e.g., phenolic aniline blue, erythrosin, acridine orange or fluorescein isothiocyanate) and the number of bacteria or length of mycelium in a series of randomly chosen microscope fields determined.

The technique is easier to carry out than the agar-film method but suffers from the fact that the smear may not dry uniformly, thus concentrating organisms on particular parts of the slide. This difficulty may be overcome by mixing the soil with a known number of readily identifiable reference particles. By counting them and calculating the ratio of reference particles to organisms, errors arising from the unevenness of drying can be corrected. Choice of reference particles depends upon the stain chosen, but indigo or minute polystyrene beads are suitable for ordinary stains, zinc cadmium sulphide for fluorescent stains (Tyldesley, 1967).

A suitable staining procedure for phenolic aniline blue is given on p. 21 and acridine orange on p. 22, while the use of fluorescein isothiocyanate is described below. Choice of stain is determined by the purpose of the experiment and the apparatus available. Phenolic aniline blue and erythrosin are good general stains that can be used for ordinary microscopy; acridine orange and fluorescein isothiocyanate can only be used in conjunction with ultra-violet illumination. Acridine orange is supposed to differentiate living (green stained) from dead (orange stained) cells, but since this difference is dependent upon cell structure and dye concentration, its accuracy is suspect. Fluorescein isothiocyanate is a protein stain and enables organisms to be distinguished from soil particles with greater ease than any of the other stains.

(i) Fluorescein isothiocyanate (Pital *et al.*, 1966; Babiuk & Paul, 1970). To prepare a staining solution, 1·3 ml of 0·5M carbonate–bicarbonate buffer (pH 9·6), 6·0 ml of 0·001M phosphate buffered saline (pH 7·2), 5·7 ml of 0·85% physiological saline are mixed and 5·3 mg of crystalline fluorescein isothiocyanate (Baltimore Biological Laboratories) added. The solution is mixed for 10 minutes.

0·1 ml aliquots of the solution are pipetted on to the air dried, heat fixed smears which are then placed in a moist chamber for 1 minute at 37°. After staining, the smears are rinsed and washed gently for 10 minutes in 0·5M carbonate–bicarbonate buffer, blotted dry, mounted in glycerol (pH 9·6, adjusted with 2% carbonate–bicarbonate buffer) and examined under ultra-violet light. Stained cells fluoresce a bright apple green.

FITC solutions only last for a few hours so that only the required amount should be made up on each occasion.

### 6.4.3 Dilution plate count

(a) *Theoretical considerations.* The dilution plate count is the most frequently used technique for determining the number of viable microbial cells in soil and, in addition, may be used as a method for isolating micro-organisms from soil (see p. 40). The technique is based upon the assumption that when a known weight of soil is agitated in a suitable liquid, the micro-organisms become detached from the soil and each of the detached cells gives rise to a discrete colony on a nutrient medium in a Petri dish. These colonies are counted and the number of cells in the original soil sample estimated. Since the number of cells even in a small soil sample is large, the suspension of cells must be diluted so that a small number of well separated colonies develop on each Petri dish.

There are a great many potential inaccuracies in this technique and it is generally agreed that the counts obtained are an underestimate of the total viable population of cells, whereas direct cell counts (see p. 61) overestimate the same population. Skinner, Jones & Mollison (1952) have discussed the reasons for this and point out that plate counts may underestimate the numbers of organisms because clumps of cells remain attached to soil particles or are aggregated together in the suspension, the dilution medium may kill cells, spores may fail to germinate, cells may be adsorbed onto the walls of the pipettes used to dilute the suspension and finally the plating medium and the incubation conditions may be highly selective.

The plate count technique is not equally useful for all groups of micro-organisms. In general, it can be said that it is only useful as a basis of biomass determinations if the propagules of the organisms being examined are of roughly uniform size and weight. Thus, it can provide useful information for both bacteria and yeasts, but is less useful for actinomycetes and filamentous fungi where the colonies that develop originate mainly from spores and cannot be distinguished from the colonies arising from hyphae (see p. 43 for a further discussion of this problem).

The outcome of the technique is affected by many factors and even the slightest change may alter the final count. For the purposes of biomass determination, any variation in procedure should be aimed at attaining the highest count possible and for this reason Parkinson *et al.* (1970) have suggested that individual laboratories may wish to vary the details of technique, rather than adopt a standard method of dubious value. Nevertheless, when the technique is used in any of its forms it should be described clearly. Therefore, the following description of the plate count technique is given not as a standard method, but to provide some idea of those variables which must be specified when using the method.

(b) *Description of the dilution plate count technique.* Take about 0·5 g of soil and add it to 500 ml quarter-strength Ringer's solution in a litre conical flask. Take another sample of soil and determine its water content and/or density and/or specific surface area. Place a $2·5 \times 0·8$ cm sterile stirrer bar in the flask and place the flask on a magnetic stirrer. Disperse the soil with the magnetic stirrer operating at about 2800 rpm for 30 minutes.

Immediately following dispersion makes a series of 10-fold dilutions of the suspension by pipetting 1 ml aliquots into 9 ml of quarter-strength Ringer's solution. It is usually sufficient to make 10-, 100- and 1000-fold dilutions, but more may be required. One ml blow-out pipettes of known tolerance should be used and prior to each pipetting operation, the liquid to be transferred should be sucked into and blown out of the pipette ten times in order to saturate the absorption sites on the pipette wall and to ensure further dispersion.

One ml aliquots of each dilution should be transferred to 7·5 cm diameter Petri dishes, in which is present a thin layer of solidified water agar (to prevent bacteria spreading); 5–10 replicate plates should be prepared at each dilution.

Twenty ml of molten medium at 45° should then be poured over the inoculum and the inoculum dispersed in the agar by oscillating the dish gently backwards and forwards, side to side, clockwise and finally anticlockwise, five times for each operation. Alternatively, surface plating (see p. 41) may be used.

After allowing the medium to set, plates should be stacked upside down in piles of six in an incubator at a constant temperature, usually between 10° and 25°. Colonies may be counted after 14 days, or at 21 days if colonies are slow to develop.

Colonies should be counted with the aid of a magnifying lens at those dilution levels at which 20–200 colonies develop on each plate. If fungi or spreading bacteria occupy more than 15% of the surface of the agar, the plates should be rejected since suppression of colony growth will almost certainly have occurred. If separate bacterial and actinomycete counts are to be made, recognizable colonies of actinomycetes should not be included in bacterial counts and *vice versa*.

The degree of replication must be recorded, including the number of replicate dilution series made, and in the case of counts of organisms in composite soil samples, the number of sub-samples used to make up the composite sample.

Useful discussions of the plate count technique will be found in papers by Brierley *et al.* (1928), Egdell *et al.* (1960), Parkinson *et al.* (1970), James & Sutherland (1939), Jensen (1962, 1968), Skinner, Jones & Mollison (1952) and Smith & Worden (1925).

### 6.4.4 Extinction dilution method

Sometimes it is not convenient or possible to use the dilution plate count for estimating numbers of viable bacteria. Often when one wishes to count the number of bacteria in a particular physiological group, no suitable solid medium may be available; the numbers of bacteria may be so low that, at the dilutions used, soil particles may interfere with colony counting. It is then useful to use the extinction dilution method in which replicate tubes of liquid media are inoculated with 1 ml aliquots of a dilution series. Growth or no growth (of a given reaction) is recorded after incubation for 14 days. The results are then referred to probability tables and the most probable number of bacteria per ml of the suspension calculated (see below). Unfortunately, the precision of the method is poor and always less accurate than the plate count for the same expenditure of effort (Taylor, 1962).

The method may be carried out with the same general precautions and procedures as the plate count. Usually five replicate tubes, each containing 6 ml of medium, are used.

Probability tables for 5, 8 and 10 tubes per dilution are to be found in Meynell & Meynell (1965). The following account of the use of the probability tables is substantially the same as that given by the above authors.

Suppose a soil suspension was diluted and 1 ml was inoculated from the $10^{-4}$, $10^{-5}$ and $10^{-6}$ dilutions into 5 tubes for each dilution. After a period of incubation, the numbers of turbid tubes at each of these dilutions were 5, 3 and 0 respectively. Table 1 shows that the most probable number of organisms is 7·9 per inoculum taken from the $10^{-4}$ dilution or $7·9 \times 10^4$ per ml of undiluted suspension. Assuming the soil suspension had been made by suspending 1·0 g of soil in 1000 ml diluent, the numbers of bacteria per g of soil would be $7·9 \times 10^7$.

## 6.5. Partial sterilization—reinoculation method (Jenkinson, 1966)

Soils which have been exposed to various partial or complete sterilization procedures are reinoculated, incubated and their output of $CO_2$ measured. All methods of sterilization render a small fraction of soil organic matter decomposable. This fraction is postulated to be the soil biomass. It has been suggested that the size of the soil biomass can be roughly estimated from the size of the flush of $CO_2$ after chloroform vapour treatment of soil.

The following procedure can be used:

Chloroform treatment. Moist soil is placed in a desiccator lined with wet filter paper and containing alcohol-free chloroform in a beaker. The desiccator is evacuated until the chloroform begins to boil and is then placed in the dark for 24 hours, at room temperature. The beaker is removed and the desiccator evacuated and opened six times to remove all chloroform vapour.

Re-inoculation. Treated soil (approx. 10 oz oven dry wt), in a glass vial, is inoculated with a 1 ml sample from an untreated soil-water suspension (1 g soil in 200 ml water). The vial is placed in a 500 ml wide-mouthed bottle with a test tube containing 5 ml of 25% tetramethylammonium hydroxide. Five ml of water is placed in the bottle which is bunged and kept at 25° in the dark for 10–20 days. An unsterilized sample of soil is treated in the same way and a blank without soil is also set up.

TABLE 1. Table of values of the most probable number for five tubes undiluted from each of three successive 10-fold dilutions (as modified from Taylor (1962) by Meynell and Meynell (1965))

| Numbers of turbid tubes observed at three successive dilutions | | | MPN (per inoculum of the first dilution) |
|---|---|---|---|
| 0 | 1 | 0 | 0·18 |
| 1 | 0 | 0 | 0·20 |
| 1 | 1 | 0 | 0·40 |
| 2 | 0 | 0 | 0·45 |
| 2 | 0 | 1 | 0·68 |
| 2 | 1 | 0 | 0·68 |
| 2 | 2 | 0 | 0·93 |
| 3 | 0 | 0 | 0·78 |
| 3 | 0 | 1 | 1·1 |
| 3 | 1 | 0 | 1·1 |
| 3 | 2 | 0 | 1·4 |
| 4 | 0 | 0 | 1·3 |
| 4 | 0 | 1 | 1·7 |
| 4 | 1 | 0 | 1·7 |
| 4 | 1 | 1 | 2·1 |
| 4 | 2 | 0 | 2·2 |
| 4 | 2 | 1 | 2·6 |
| 4 | 3 | 0 | 2·7 |
| 5 | 0 | 0 | 2·3 |
| 5 | 0 | 1 | 3·1 |
| 5 | 1 | 0 | 3·3 |
| 5 | 1 | 1 | 4·6 |
| 5 | 2 | 0 | 4·9 |
| 5 | 2 | 1 | 7·0 |
| 5 | 2 | 2 | 9·5 |
| 5 | 3 | 0 | 7·9 |
| 5 | 3 | 1 | 11·0 |
| 5 | 3 | 2 | 14·0 |
| 5 | 4 | 0 | 13·0 |
| 5 | 4 | 1 | 17·0 |
| 5 | 4 | 2 | 22·0 |
| 5 | 4 | 3 | 28·0 |
| 5 | 5 | 0 | 24·0 |
| 5 | 5 | 1 | 35·0 |
| 5 | 5 | 2 | 54·0 |
| 5 | 5 | 3 | 92·0 |
| 5 | 5 | 4 | 160·0 |

Determination of $CO_2$. $CO_2$ absorbed by the tetramethylammonium hydroxide is determined as follows. The tetramethylammonium solution is made up to 10 ml with boiled-out distilled water. A 3 ml aliquot is placed in a 100 ml beaker on a magnetic stirrer and 25 ml of boiled-out distilled water added. The pH of the solution is brought to 10·0 by slow addition of $NH_4Cl$ and then to 8·3 by addition of 0·05N HCl. The amount of $CO_2$ evolved is calculated from the volume of 0·05N HCl needed to bring the pH from 8·3 to 3·7, less that required by the blank.

Estimation of biomass. This is calculated from the formula

$$B = \frac{F}{K}$$

where $B$ = amount of carbon in biomass,

$F$ = flush of decomposition, i.e., the difference between $CO_2$ evolved from chloroform treated soil and that from untreated soil, after the initial disturbance effect has passed. It is recommended that the $CO_2$ evolved from untreated soil in the second 10 days of the 20-day incubation period is used,

$K$ = % of biomass carbon mineralized to $CO_2$. This was estimated from experiments with dead bacterial cells to be 30% in 10 days at 25°. Therefore, $K = 0·3$.

e.g., $CO_2$ evolved in 1st 10 days after chloroform treatment = 5·11 mg carbon/100 g soil.

$CO_2$ evolved in 2nd 10 days from untreated soil = 0·53 carbon/100 g soil

$$F = 4·58$$

$$B = \frac{4·58}{0·3}$$

$$= 15 \text{ mg carbon/100 g soil.}$$

## 6.6 Chemical estimations

Recently there have been attempts to utilize quantitative chemical analyses of soil for organic compounds unique to specific groups of soil micro-organisms. The data obtained have been used to provide estimates of the microbial populations in these soils. It is too early to say whether such methods can provide accurate measures of biomass but they are of sufficient interest and promise to be described.

F

### 6.6.1 Muramic acid content of soil (Miller & Casida, 1970b)

Mild alkaline treatment of muramic acid results in the quantitative release of D-lactic acid. Assay of this and comparison of the results with those for pure cultures of soil bacteria yield measures of bacterial numbers in soil. The procedure is as follows.

Ten gram samples of soil are hydrolysed in 200 ml boiling, 6N hydrochloric acid for 4 hours in a reflux column. The acid is then removed by evaporation *in vacuo* at 46–48° and the resulting residue suspended in distilled water. After centrifugation, the supernatant is made up to 50 ml with distilled water and stored at $-10°$.

Five ml portions are neutralized with 10 ml of 1·0M sodium bicarbonate solution, centrifuged and the clear supernatant acidified with 3·5 ml 2·5N acetic acid. This solution is passed through a Dowex 50W–X2, 200–400 mesh, cation exchange resin prepared in the hydrogen form and packed as a 10 cm bed in a $1 \times 25$ cm glass column. Two or more columns are used for each soil sample to prevent overloading. After adsorption, each column is washed with 10 ml water and eluted with 10 ml of 2N hydrochloric acid. The various eluates are combined and the acid evaporated *in vacuo*.

The resulting amino compounds are resuspended in 5·0 ml distilled water and added to 4·5 ml 0·08M sodium phosphate adjusted to pH 12·5 with concentrated KOH. After incubation for 2 hours at 37°, the samples are acidified with 0·5 ml 12N hydrochloric acid and passed through a cation exchange column as described above. The first 6 ml to pass through the column is discarded but the net 4 ml is analysed for lactic acid (Markus, 1950). Each ml collected contains the lactic acid released from the muramic acid present in 0·1 g of the original sample.

Average levels of muramic acid in pure cultures of bacteria were found by Millar & Casida to be 0·004 $\mu$g per million Gram-positive bacteria, 0·0005 $\mu$g per million Gram-negative bacteria and 0·066 $\mu$g per million aerobic bacterial spores. However, levels in soil were found to be much higher than could be accounted for by these results, i.e., about 5·6 $\mu$g per million platable cells. This suggests that there are many more muramic acid-containing bacteria in soil than are found on dilution plates, or that bacteria in soil accumulate muramic acid as a storage product or that there are dead cells contributing to the figure obtained for soil. An inability to distinguish between these possibilities prevents the adoption of this method for accurate measurement of the bacterial biomass of soil.

# 7

# Determination of Microbial Activity in Soil

## 7.1 Introduction

Biological activity has been measured in a number of ways. Some workers have attempted measurements of *actual* biological activity in undisturbed soil in their field sites, whereas others have removed soil from its natural setting into the laboratory and there measured the *potential* activity of the soil biota under defined conditions (e.g., temperature, nutrient status, moisture regime, $O_2/CO_2$ regime). Differences in potential activity between soil samples may reflect differences in the actual activity in the sites from which they were drawn, but this is not necessarily the case.

In the strictest sense biological activity can only be measured in metabolic terms, e g., oxygen uptake, carbon dioxide evolution, enzyme activity or heat evolution. However, such measurements do not usually allow one to distinguish the relative contribution of the different living components of soil, i.e., bacteria, fungi, animals and plant roots, to the total activity because the metabolic processes measured are universal.

It has been suggested that the contribution of an individual group of organisms could be determined by selectively inhibiting it and measuring the residual rate of metabolism. However, it is difficult to bring inhibitors into contact with sensitive organisms in soil microhabitats, the inhibitors applied and/or the dead remains of organisms sensitive to the inhibitor may act as substrates for the remaining organisms, and the removal of one group of soil organisms may release other groups from their competitive influences. It is more usual, therefore, to consider individual soil processes rather than individual groups of organisms. Thus rate of substrate transformation can be used as an indication of activity, e.g., cellulose breakdown, ammonification.

Despite the foregoing comments many soil microbiologists have attempted to infer biological activity of major groups of soil micro-organisms from data obtained in other ways, e.g., presence or absence on isolation plates, biomass. These inferential measurements are of little value when one is interested in decomposition processes and energy flow. However determination of the ratio dormant propagules : vegetative cells gives a clue to the level of activity in a soil sample.

Whilst the ultimate aims of activity studies are to determine rates of microbial growth (productivity) and energy dissipation due to micro-organisms it is clear from the foregoing that this is not yet possible. Data has been obtained on the growth and respiratory activity of pure or mixed cultures of soil organisms in sterile soil or soil-like systems. This type of study may allow the development of concepts related to productivity even though they are not immediately relevant to field situations because of the absence of antagonistic and synergistic interactions.

## 7.2   Measurement of metabolic activity

### 7.2.1   Field methods for measuring gas exchange

Whilst $CO_2$ evolution rate measurement has been most commonly applied for assessing biological activity in soil samples or in undisturbed soil, such measurements have several difficulties. The conditions imposed by the apparatus may cause deviations from normal biological activity, underestimates of $CO_2$ evolution may be caused by carbon compounds other than $CO_2$ being released to the atmosphere or to the ground water and overestimates of soil microbiological activity may be made as a result of $CO_2$ evolved by root and soil animal respiration.

(a) *Inverted box method for determination of $CO_2$ evolution (Witkamp, 1966).* Round boxes each of 175 cm² surface area are inverted and the rims sunk 2·5 cm into the otherwise undisturbed field soil. A dish containing 5 ml of 0·1N KOH is placed in each box, and a one hour equilibration period is allowed. Amounts of $CO_2$ evolved are measured by titration using 0·179N HCl (Conway, 1950).

(b) *Method of Wallis & Wilde (1957) for determination of $CO_2$ evolution.* Air is extracted from soil through metal cylinders or glass funnels, 12·75 cm diameter, firmly embedded in the soil. Extraction of air from the enclosed soil is by means of a battery-operated pump (e.g., Trico 'Electro Vac' pump operated by a six volt battery).

Following extraction from the soil, the air is drawn through N NaOH solution held in a test tube, then through a safety bottle to the pump. A 'T' valve placed between the safety bottle and the pump allows control of air flow rate. Rubber tubing with capillary tips is used to connect the components of the apparatus.

Carbon dioxide in the air drawn from the soil is absorbed by the N NaOH for a period of 2 hours, the $Na_2CO_3$ formed treated with an excess of 2N $BaCl_2$ and residual NaOH titrated with N HCl. Hence the volume of NaOH converted to $Na_2CO_3$, and thus the amount of $CO_2$ absorbed, can be calculated:

vol. NaOH converted (to $Na_2CO_3$) $\times 0\cdot022$ = wt. (g) of $CO_2$ absorbed (i.e. wt. of $CO_2$ evolved from the enclosed soil).

These data can be converted to the amount of $CO_2$ evolved per square metre of soil surface by multiplying the figure, obtained above, by:

$$\frac{10,000}{\pi r^2}$$

where $r$ = radius (in cm) of the metal cylinder or glass funnel.

(c) *Method described by Reiners (1968).* This is a more sophisticated method of monitoring $CO_2$ evolution from soils.

A steel sleeve (20 cm × 50 cm × 20 cm) is inserted into the soil until the top edges are slightly above the top of the litter layer. The sleeve is covered with a tightly fitting plastic lid which is provided with an inflow and an outflow tube. Ambient air is pumped into the enclosed system at measured rates of about 19–25 litres/min, distributed across the 20 cm width of the enclosed soil by a manifold and passed through the outflow tube. Air being pumped into and withdrawn from the apparatus is sampled for $CO_2$ content with an infra-red gas analyser (e.g., Beckman, model 15-A) while air flow rates are measured with a Brooks rotameter and air temperature with a Tri-R electronic thermometer. Replicate steel sleeves are placed in each soil being studied.

### 7.2.2 Laboratory methods for measuring gas exchange

(a) *Carbon dioxide evolution*

(i) Static systems.

Ink Bottle Technique (Kibble, 1966). Soil samples are air-dried overnight, sieved (if necessary to remove large particles) and 10–20 g sub-samples placed in screw cap ink bottles which have a well in the neck (see Fig. 5). If the soil is to be amended, it is done at this stage.

A pre-determined amount of water is added to bring the soil to a known moisture content, or preferably a known pF status, mixed thoroughly and tamped down. Three ml of 10% NaOH is placed in the well of the bottle. The bottle top is greased, capped and incubated at the appropriate tempera-

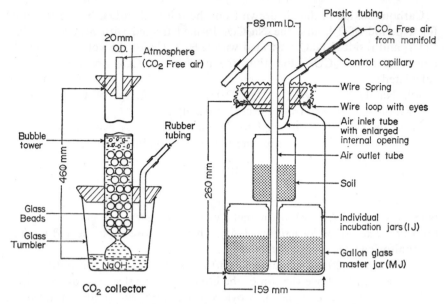

Figure 5. Incubation unit for measuring $CO_2$ evolution from soil. The unit shown is used when sub-samples of soil are removed during incubation. When this is not required, soil is placed directly in the master jar. From Stotzky G. (1965) in *Methods of Soil Analysis*, II, ed. C. A. Black *et al*. American Society of Agronomy, Madison, Wisconsin.

ture. Suitable controls must be included to allow for calculation of $CO_2$ present in the bottles.

The alkali is removed from the bottles at appropriate intervals, depending on the degree of microbial activity, and the determination of $CO_2$ absorption by the alkali made by the titration method described earlier (see page 73).

After titration, the bottles are left open to the air to equilibrate for one-half hour before further use.

This technique does not involve the forced aeration of the soil samples and the air within the reaction vessel is changed only when the screw caps are removed. It is cheap to use and has been found valuable for initial studies on the effects of different soil amendments on soil microbial activity.

(ii) Aerated systems. Various techniques have been described for the measurement of $CO_2$ evolution from periodically or continuously aerated soil samples. An example of this type of technique has been given by Stotzky (1965) and the apparatus for this technique is shown in Fig. 5.

Weighed quantities of soil are placed in incubation jars. The air inlet tube is attached to the manifold which is in turn attached to the scrubber system (to remove $CO_2$). A constant flow of air over the soil sample(s) is maintained at 10–15 litres/hr.

Residual air is flushed from the vessels and the air outlet tube attached to a vessel containing a known amount of standard KOH or NaOH solution (of a strength so that not more than two thirds is neutralized by $CO_2$). As a control for $CO_2$ absorbed from the atmosphere, $CO_2$ collectors are attached to empty incubation vessels and replaced periodically. When determining the $CO_2$ absorbed, each bubble tower is rinsed into the alkali using $CO_2$ free water.

An excess of $BaCl_2$ is added to precipitate $BaCO_3$ and the unneutralized alkali titrated with standard HCl in the container using phenolphthalein as indicator, or without phenolphthalein using an automatic titrator.

One criticism of this technique is that the inflowing $CO_2$-free air only sweeps the surface of the soil samples, but it is easy to modify the equipment so that $CO_2$-free air passes through the soil sample. It has been demonstrated that the rate of air flow may have important effects on total microbial activity.

As well as titrimetry, numerous other methods for the analysis of $CO_2$ concentrations in gas samples (obtained from the aeration of soil under experiment) have been used. These include the Haldane gas analyser (Baver, 1948), infra-red absorption (Miller, 1953), gas chromatography (Keulemans, 1959) and mass spectrometry (Dunning, 1955).

### (b) *Oxygen uptake measurement*

(i) Warburg technique (Katznelson & Rouatt, 1957). The application of this technique allows measurements of $O_2$ uptake by soil samples over short periods of time (i.e., a few hours) as well as over long experimental periods. The technique described here is based on that outlined by Katznelson & Rouatt (1957).

Samples of soil of known moisture holding capacity are quickly air-dried using a fan and sieved. Soil particles greater than 1 mm in diameter are rejected. Four gram samples of the sieved soil are placed in each Warburg flask and water is added to the soil in each flask to bring the soil to 70 % of its moisture holding capacity. At this stage it has been found by various workers that it is necessary to allow the soil to settle in the flasks for a period of 16–20 hours before proceeding. If this is not done, erratic and high $O_2$ uptake readings are obtained in the first period of the determinations.

After the settling period, 0·2 ml 20 % KOH is carefully added to the centre well of the Warburg flask to absorb $CO_2$ evolved from the soil sample. A piece of fluted filter paper placed in the centre well prior to the addition of KOH solutions aids $CO_2$ absorption.

The flask is attached to a manometer and placed in a water bath at the required temperature. A period of temperature equilibration is necessary, during which time the manometer is open. After this equilibration period the manometer is closed and readings begun at intervals which depend upon the degree of microbial activity.

The application of this technique for tissue respiration and its principles are thoroughly described by Umbreit, Burris & Stauffer (1964) who explain the theoretical reasoning involved in the calculation of $O_2$ uptake data from manometer-level changes. A summary of the calculations involved for the application of this method for measuring $O_2$ uptake of soil samples is given below:

$$\mu l \text{ Oxygen taken up} = hK_{O_2}$$

where $h$ is the difference (in mm) between original and final heights of fluid in the open arm of the manometer, corrected for change in atmospheric pressure by adding or subtracting the fluid level changes in in the open arm of a thermobarometer.

$K_{O_2}$, the flask constant, is calculated by a modification of the formula derived by Umbreit, Burris & Stauffer (1964):

$$K_{O_2} = \frac{V_g\, 273/T + V_f\, \alpha\, O_2}{P_0}$$

where $V_g$: volume of gas space
$\quad\quad V_f$: volume of fluid in the flask
$\quad\quad \alpha$: solubility in reaction liquid of the gas involved (in this case oxygen) when the gas is at a pressure of 760 mmHg at the temperature $T$.
$\quad\quad T$: experimental temperature (° Absolute)
$\quad\quad P_0$: standard pressure, 760 mmHg = 10,000 mm Brodie's fluid or Kreb's fluid
$\quad\quad V$: volume of flask and manometer down to the calibration mark on the manometer.

In the liquid-gas systems described by Umbreit *et al.*

$$V_g = V - V_f$$

When using soil in Warburg systems

$$V_g = V - V_f - V_s$$

where $V_s$ : volume of soil in the flask, and the formula for calculation of the $K_{O_2}$ per g oven dry soil now reads:

$$K_{O_2}/\text{g oven dry soil} = \frac{(V - V_f - V_s)\,273/T + V_f\,\alpha\,O_2}{P_0 \times \text{oven dry wt of soil}}$$

This technique has been criticized by some soil ecologists on the grounds that:

1. The preparation of the soil prior to incorporation within the Warburg flask causes gross disturbance of the soil fabric.

2. Frequently the temperatures, under which soil respiration are measured, are high, e.g. 25°C, in relation to the temperature of the soil from which the samples were taken.

3. Measurements of $O_2$ uptake are undertaken in atmospheres which are $CO_2$-free, a factor which may induce changed microbial activity as compared with that observed in soil in the field.

It is appreciated by many workers who use this technique that the data obtained do not allow calculation of the actual microbial activity in natural undisturbed soil horizons but only a comparison of the potential activity of soil samples from different soils and different horizons of the same soil. One of the main advantages of the method is that it allows the accurate measurement of $O_2$ uptake from small replicate soil samples over short periods of time i.e., a few hours.

The technique can be extended in various ways: e.g.,

1. To study the effect of inhibitors of specific components of the soil microflora or fauna (although in this type of work it is frequently difficult to ensure that the inhibitor reaches the exact sites of biological activity within the soil sample—see also p. 71).

2. To study the effects of substrates (e.g., sugars, cellulose, lignin, etc.) added to the soil samples and thus obtain data on the potential biochemical activities of the soil microflora.

3. To study the effects of various environmental conditions on soil microbial activity, since such factors as temperature, moisture content and $CO_2$ concentration within the flasks can be controlled.

(ii) Modifications of Warburg Technique (Parkinson & Coups, 1963). This technique allows measurement of oxygen uptake of either air-dried re-moistened soil or fresh untreated soil with a larger soil sample than in conventional Warburg flasks, i.e., up to 25 g.

Soil or litter samples (the weight used depending on the specific gravity of the material under study) are collected from the soil horizon under study and stored at 2°C overnight. Large particles such as twigs and stones, are removed by hand and sieving of some samples may be necessary. Measurements of moisture content and organic matter content of the soil are made.

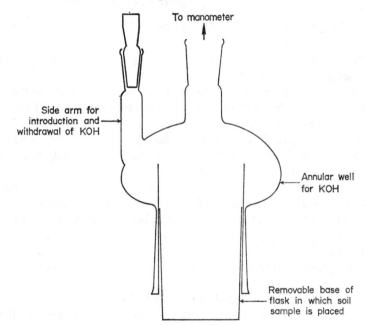

Figure 6. Diagrammatic section of a respiration flask used for measurement of $O_2$ uptake of $F_2$–C horizon material. From Parkinson D. & Coups, E. (1963) in *Soil Organisms*, ed. J. Doeksen & J. van der Drift. North Holland Publishing Co., Amsterdam.

Five to thirty-five g sub-samples are placed in the removable bases of respiration flasks shown in Fig. 6, and the flask is then assembled. A manometer (tap open) is attached to the flask and the system placed in a water bath at the desired temperature for study. The soil is allowed to settle and equilibrate for 20 hours before 10% KOH is placed in the annular well of the flask via the side arm and the manometer tap is closed. Hourly readings of oxygen uptake by the soil over a 5-hour period (or longer periods) are made.

Calculations of data (in terms of $\mu l$ $O_2$ taken up/g dry wt. soil) are made as for the conventional Warburg technique.

This modification of the Warburg method is susceptible to all the criticisms which have been levelled at the conventional application of this method. The advantages of the modification described are:

1. Larger samples of material can be dealt with which produces more consistent results for samples from the same soil horizon or samples exposed to the same soil treatment.

2. The KOH can be readily replaced in the annular well via the side arm which facilitates long-term experimentation. Also titration of the alkali is possible, allowing the determination of $CO_2$ evolution from the soil.

3. The soil sample is easily placed in and removed from the flask, allowing easy determination of chemical and/or biological properties of the soil whose activity has been measured.

(iii) Study of the effects of soil amendments by the Warburg technique. It has been mentioned previously that the Warburg technique can be used for the study of the effects on microbial activity of both inhibitory and stimulatory amendments to soil samples. The techniques used for such studies are varied but frequently the procedure adopted is of the following type.

Before placing the soil in the flask, an air-dried sample is divided into sub-samples, the number of sub-samples depending on the number of soil treatments to be studied. Each sub-sample is amended with a specific concentration of one of the materials under study, allowing for proper control sub-samples. Weighed amounts of the amended and control soil samples are placed in separate, replicated Warburg flasks and $O_2$ uptake measurements made as described earlier (p. 75).

In studies on the effects of environmental conditions, e.g., temperature, moisture, etc. on microbial activity in mineral horizons of soil where very low rates of actual activity are found, it is frequently difficult to obtain data which is amenable to proper statistical analysis. It is therefore necessary to amend the soil samples to allow high rates of microbial activity and to allow observation of larger differences in such activity brought about by varying the environmental conditions. Whether this procedure is justified or not has rarely been considered.

(iv) Oxygen uptake measurements on undisturbed cores of soil and litter (Howard, 1968). Weighed amounts of soil are placed in glass tubes (28 mm diameter $\times$ 15 cm length) closed at one end with glass wool. Known amounts of fragmented litter are placed on top of the soil or mixed with it and kept moist.

The tubes are placed upright in a box in the field and protected from rain by placing a roof over the box. The tubes are removed from the field at various intervals for oxygen uptake measurements and kept in the laboratory at temperatures similar to those in the field.

To measure oxygen uptake the tubes are placed in the base of a flask (a 'Quickfit' B 40 cone with a flat base 15 cm from the ground end), along with approximately 5 mm of water to ensure gas exchange occurring only at the top of the tube. 1N NaOH is placed in the well of the flask and the flask connected to a Dixon manometer.

A minimum of 10 hours equilibration in the water bath at a temperature corresponding to field conditions is suggested before manometric measurements are begun.

The advantages of this method are:

1. The same soil and litter tubes can be used for repeated measurements.

2. Removing the sample from the field does not involve gross disturbance of the sample.

3. Disappearance of the litter other than by decomposition does not occur.

4. The environmental conditions influencing the samples closely resemble those in the field.

(v) Recently the Gilson differential respirometer (Umbreit *et al.*, 1964) has come into use in studies of soil and litter respiration. This apparatus is a modification of the Dixon constant pressure respirometer in which there is direct measurement of changes in gas volume. However, instead of each manometer having a compensating flask of exactly the same volume as that of the reaction flask and in which is placed inactive material (as in the Dixon system) all the manometers are linked to a common compensating flask. The use of the Dixon & Gilson respirometers in soil and litter studies have been discussed in detail by Howard (1968).

Stotzky (1965) considers that, whilst oxygen uptake and carbon dioxide evolution have both been used frequently to measure 'soil respiration', carbon dioxide evolution is more appropriate for such studies. The reasons advanced (Stotzky, 1965) for this conclusion are:

1. In the manometric determination of oxygen uptake, gases other than $CO_2$ may be evolved as a result of microbial activity and may interfere with the manometric measurement.

2. In using oxygen uptake measurements to assess microbial activity an R.Q. of 1 is assumed—a condition seldom fulfilled in soil.

3. For oxygen uptake to be an accurate reflection of soil respiratory activity, the soil environment must be completely aerobic. In anaerobic situations, which occur in some soil micro-habitats, carbon dioxide is evolved without oxygen uptake. Thus, in soil samples, it is likely that oxygen uptake measurements underestimate the level of activity.

4. As with carbon dioxide, non-biological factors may interfere with the use of oxygen uptake measurements as indices of microbial activity.

However, Stotzky (1965) points out that there are also problems in use of $CO_2$ evolution measurements as indices of biological activity. These stem mainly from non-biological production of $CO_2$ through chemical decarboxylation, cell-free enzymes, or from free carbonates in soil.

In assessing the respiratory quotient, other factors affecting $CO_2$ evolution must be considered, i.e., decarboxylase activity (yielding $CO_2$ without $O_2$ uptake) and $CO_2$ fixation. Stotzky (1965) points out that with the mixed populations found in soil such factors are probably of minor importance.

### 7.2.3 Enzyme assay methods

Although a wide range of enzymes in soil have been studied, e.g., saccharase, invertase, maltase, amylase, xylanase, $\beta$-glucuronidase, asparaginase, proteinase, urease, phytase, dehydrogenase, catalase, and glycerophosphatase, only methods for assaying dehydrogenase and urease activity, both of which have been used widely in investigations of microbial activity in soil, will be considered.

    (i) Dehydrogenase activity (Lenhard, 1956). Twenty g fresh soil is mixed in a 50 ml beaker with 200 mg dry $CaCO_3$ and brought to 90% water holding capacity with water containing 2·0 ml of a 1% solution of 2,3,5-triphenyltetrazolium chloride (TTC). The soil is thoroughly mixed and the surface tamped to prevent access of air. If dehydrogenase activity is present in the soil, TTC is reduced to triphenylformazan (a red coloured compound).

$$2,3,5\text{-triphenyltetrazolium} + 2H \rightarrow \text{triphenylformazan} + HCl$$
$$\text{(TTC)} \qquad\qquad\qquad \text{(TPF)}$$

The sample is incubated at 30°C and 70% relative humidity for 24 hours. Then 25 ml concentrated methanol is added and stirred for 5 minutes. The resulting slurry is washed into a Buchner funnel containing a Whatman No. 5 paper and extracted with successive aliquots of concentrated methanol. The volume of extractant used for individual samples is recorded and the density of the extracted coloured triphenylformazan determined spectrophotometrically at a wavelength of 546 nm, using concentrated methanol as reference blank. Concentrations of the formazan in the extract are calculated by comparison with a standard curve of triphenylformazan (TPF) in methanol.

Results can be recorded in volumes of hydrogen transferred during reduction of TTC to TPF in 20 g soil. The formation of 1 mg TPF requires 150·35 $\mu$l hydrogen.

Kibble (1966) examined various aspects of Lenhard's method (1956) and described the following modified technique.

Twenty g fresh soil plus 1 ml of 3 mg/ml MTT (dimethylthiazolyl tetrazolium bromide) solution are mixed in a beaker. The addition of $CaCO_3$ was suggested by Lenhard, 1956, to counteract the inhibitory effect of low pH on microbial multiplication. If $CaCO_3$ is added then tests must be made on the effects of such an addition.

The mixture is incubated at 25°C at 60% moisture holding capacity under nitrogen for 6 hours in an anaerobic jar. The shorter incubation time and lower concentration of tetrazolium is used to decrease the adverse effects of formazan formation on microbial activity.

After 6 hours, benzene is added to the samples and formazan extracted by successive freezing. Beakers are immersed in a salt water freezing mixture in the freezing compartment of a refrigerator for 10 minutes and thawed at room temperature. Benzene, being a non-polar solvent, extracts least humic material from soil and is as efficient as other solvents, e.g., methanol and acetone, in extracting dye from bacteria and actinomycetes. For fungi, benzene is the most efficient solvent and MTT formazan the most readily extracted.

The extracted formazan solution is rendered optically clear by centrifuging, and the optical density is measured at 560 nm. Controls are set up to allow compensation for any extracted soluble organic matter. The amount of formazan in solution is calculated from a calibration curve of formazan in benzene. The formation of 1 mg of formazan requires 66·64 $\mu$l hydrogen.

The use of dehydrogenase assay has not been considered a useful quantitative method for assessing metabolic activity of micro-organisms in soil because it is reputed to have less than 5% of the efficiency of oxygen uptake measurements. However, some workers have given data which indicate that this method can provide rough comparative estimates of microbial activity in different soils. If it is desired to use this method it is essential that the technique be tested on the soil under experiment prior to its use.

(ii)  Urease activity.

(a) *Method !described by McGarity & Myers (1967).* Twenty g soil, passed through a 3 mm sieve, is placed in a 100 ml volumetric flask, 2 ml toluene added and allowed to stand for 15 minutes to permit complete penetration into the soil. Then 20 ml potassium citrate–citric acid buffer (pH 6·7) and 10 ml 10% urea are added. The flask is shaken and incubated at 37°C for 3 hours; however, lengthening the incubation period up to 6 hours may be desired. Controls, where urea is replaced by 10 ml distilled water, are run for each soil sample.

After incubation, the flask contents are made up to 100 ml with distilled water, the toluene forming a layer above the graduation mark. The flasks are thoroughly shaken, and their contents filtered through Whatman No. 5 paper. The filtrate varies from colourless to brown, depending on amount of organic matter in the soil. The presence of this colour in extracts is accounted for in subsequent colorimetric measurements by reference to control measurements.

Ammonia released as a result of urease activity is determined by the indophenol blue methods:

One ml of filtrate is placed in a 50 ml flask and made up to 10 ml with distilled water. Five ml phenolate solution (see footnote*) and 3 ml sodium hypochlorite solution, containing 0·9% active chlorine, are added, the whole mixture thoroughly mixed, and after 20 minutes made up to 50 ml.

Optical density is measured, spectrophotometrically, at 630 nm within 60 minutes. The amounts of ammonia-N formed are calculated by reference to a calibration curve. One ml filtrate corresponds to 200 mg soil. Urease activity is given per 100 g soil, i.e., number of mg ammonia-N split from urea by 100 g soil is the urease value. In the preparation of the calibration curve the same procedure is followed as described above using a calibrating solution which is made up as follows:

* Phenolate solution: The components are phenol solution and caustic soda solution.

### a.  phenol solution:

62·5 g phenol dissolved in a minimum volume of methanol denatured alcohol, 18·5 ml acetone added and the mixture made up to 100 ml with ethyl alcohol.
Keep in refrigerator.

### b.  caustic soda solution:

27 g NaOH dissolved in 100 ml water.
Keep in refrigerator.
Phenolate solution—mix 20 ml phenol solution with 20 ml caustic soda solution, dilute to to 100ml with distilled water. Use immediately.

Dissolve 4·717 g ammonium sulphate in 1 litre water. Take 10 ml of this solution and dilute to 1 litre. In the final solution,

    1 ml of solution contains 10 $\mu$g ammonia-N

    2 ml of solution contains 20 $\mu$g ammonia-N

    10 ml of solution contains 100 $\mu$g ammonia-N.

(b) *Method described by Porter (1965).* A 50 g sample of soil is placed in each of four 250 ml Erlenmeyer flasks and the flasks plugged with cotton wool. Flask 1 is sterilized and 5·0 ml urea solution (0·8 g recrystallized urea dissolved in 100 ml water and filtered through a millipore filter are added. Flasks 2 and 3 are kept non-sterile and urea solution added (as for Flask No. 1). Flask 4 is used as the control, filtered water only being added.

If it is desired to study activity under particular conditions of moisture tension it is at this point that appropriate amounts of water are added to the soil samples.

The soil samples are incubated at 30°C for 24 hours and then the control sample, the sterilized sample and one non-sterile sample are removed. The other non-sterile soil sample is left in case an incubation period more than 24 hours is necessary.

Knowing the amount of water previously added to the soil samples, saturated $CaSO_4$ solution is added to bring the final liquid volume in each flask to 100 ml. The flasks are stoppered and shaken mechanically for $\frac{1}{2}$ hour. The soil is allowed to settle and the suspensions centrifuged to obtain a clear supernatant.

Fifteen ml aliquots of clear supernatant are pipetted into 25 ml volumetric flasks and 10 ml of coloured reagent (2·0 g p-dimethylamino-benzaldehyde in 100 ml 95% ethyl alcohol and 10 ml concentrated HCl) added. The solutions are mixed and the flasks placed in a water bath at 25°C for 10 minutes.

The transmittance of solutions is read on a spectrophotometer at 420 nm, maintaining the coloured solutions at 25°C until they are ready. The extent of urea hydrolysis is calculated using a standard curve prepared for urea determinations in the usual way and the results expressed as ppm urea hydrolysed per gram soil per unit time.

### 7.2.4 Thermal measurements

Although determinations of soil respiration have been widely used as an index of microbial activity, measurement of heat evolved during organic matter decomposition provides an alternative index of activity.

(a) *Method of Newman & Norman (1943).* Air-dry soil samples are spread in a thin layer and for 7 days are exposed to an atmosphere of high relative humidity to reduce the occurrence of a heat of wetting phenomenon, a non-biological cause of thermal change which accompanies wetting of air-dried soil samples. A 50–350 g soil sample is placed in a Dewar flask (internal dimensions 7 cm × 15 cm), and an appropriate volume of water, or of solution containing the substrate under study, is added. Alternatively fresh moist soil may be used and placed into the Dewar flask without pretreatment. The flasks are kept in a constant temperature room.

Measurement of temperature changes in the soil samples are made electrically with a sensitive galvanometer, the deflection of which is a measure of the difference in e.m.f. between two thermocouples. One of the thermocouples is maintained at a constant temperature as a reference, e.g., water in a sealed Dewar flask in equilibrium with room temperature. Copper-constantin thermocouples of a 30 gauge wire are enclosed in thin-walled glass tubing for insertion into the soil samples and into the reference liquid. One thermocouple lead is led to a multiple switch to allow consecutive readings of a number of samples.

The equipment suggested by Newman & Norman, 1943, was: a Leeds and Northrup HS galvanometer of low resistance, critical resistance 18 ohms, period 5·8 second, total resistance 15 ohms. Incorporation of appropriate resistances in the circuit allows the scale deflection to be adjusted to any expected temperature differences. The galvanometer is damped sufficiently to allow stable readings to be taken quickly after deflection. The insertion of a 40 ohm resistance in parallel and a 150 ohm resistance in series was found desirable. Calibration of the galvanometer is by means of a Beckman thermometer. Using this type of galvanometer it was found impossible to secure a stable zero point after large deflections. Therefore, temperature changes were determined potentiometrically, using the galvanometer as a null point instrument and a Leeds and Northrup K type instrument with a melting ice cold-junction as a reference.

The thermo-electric power of a Copper-constantin thermocouple in the range of $20° - 30°C$ is 39·6 microvolts per degree. The system was found to be satisfactory and sensitive (Newman & Norman, 1943).

(b) *Method of Clark, Jackson & Gardner (1962)*. The instrumentation required includes temperature sensing devices, a 52-point stepping switch, clock mechanisms, bridge-type strip-chart recorder and a constant temperature room.

The temperature sensing devices quoted are Veco 32A1 glass encased thermistors (Victory Engineering Co., Union, New Jersey) whose resistance

G

at 25°C is 2000 ($\pm$ 20%) ohms. Each thermistor is brought to a common resistance of approximately 2470 ohms at 23°C by adding a suitable series resistance. This is necessary because a resistance change of 400 ohms (about 4°C) would cause a full-scale deflection of the recorder.

Forty-eight thermistors are used, one leg of each thermistor being connected to a common lead going to the recorder. The other leg is connected in series with a suitable resistance and then to the stepping switch (Switch No. R–6210, Relay Sales Inc., Box 186, West Chicago, Illinois). One lead from the stepping switch goes directly to the recorder (i.e., one thermistor at a time is connected with the recorder).

The stepping switch is energized by the clock mechanism which allows the resistance of each thermistor to be recorded for 1 minute. This time interval is chosen to allow a slight self-heating of the thermistor to equilibrate with the surroundings. Since 48 thermistors are used the remaining points on the switch are connected to a constant resistance, which serve as a check on the equipment.

The clock mechanism includes 2 synchronous motors (Cramer Controls Co., Centerbrook, Conn.) each with a plexiglass cam and a micro-switch. One motor (the switching motor) controls the stepping switch. The second motor (the control motor) controls the switching motor, the chart drive and the point-mechanism of the recorder. When the cam on the control motor activates the micro-switch, the switching motor is turned on and recording begins. When the last position of the stepping switch is reached, a shut-off relay is energized which maintains the switching motor circuit open until the control motor again closes the circuit. The relay and control motors are connected in parallel. When the control motor closes the switching motor circuit to begin the cycle, the shut-off relay is de-energized until the cycle is completed.

A one-sixth rph control motor is used, the cam being cut so that the micro-switch is activated once per revolution. Hence a recording cycle is initiated every 6 hours. Other time intervals can be obtained using motors of other speeds or by cutting the plexiglass cams in such a manner as to activate the micro-switch several times during one revolution. This is also true of the switching motor and the cam.

A Brown bridge-type strip chart recorder is used as the recording instrument. The scale division readings of the chart may be converted to temperatures by use of calibration curves, obtained by immersing the thermistors in a precisely controlled constant-temperature water bath; chart scale readings for each thermistor are obtained at various known temperatures. Scale readings are plotted against temperature. A separate calibration curve is required for each thermistor.

Soil samples, amended or unamended, are held in insulated containers. Thermistors are placed in these containers which are then covered with moisture resistant polyethylene to reduce moisture evaporation during the incubation period (30–40 days).

### 7.2.5 Rate of substrate disappearance or metabolic change

(a) *The limiting dilution method (Pochon, 1957).* Pochon has described a method for determining the potential activity of microbial populations in respect of metabolic changes detected in culture. The method has wide applications and can be used to study the appearance or disappearance of most nutrients and metabolites that can be chemically identified in a simple manner.

The method is essentially a modification of the most probable number method, although the results are expressed differently. The following description of the method is modified from that given by Pochon (1957).

One g of fresh soil is ground in a pestle and mortar with 10 ml water to give a $10^{-1}$ dilution. A series of 10-fold dilutions are prepared from this (see p. 65) and three replicate tubes containing 22·5 ml selective medium are inoculated with 2·5 ml amounts of each dilution. The tubes are incubated at the desired temperature for a suitable period of time.

At daily intervals (for 6 days) and at two daily intervals up to the end of two weeks, 2·5 ml of culture from each tube is removed. This is used for detection of metabolic changes and the following example is that given for detection of ammonification of tyrosine.

Two ml of the culture is placed in a Kahn tube and tested with four drops of Millon's reagent and acetic acid (10 drops). If a colour develops at 40°C in $\frac{1}{2}$ hour, tyrosine is still present.

0·5 ml of the culture is tested with Nessler's reagent for the presence of ammonia.

Each day the tubes showing metabolic changes are noted. The highest dilution showing change is plotted against time and the curves for different soils are compared. At the end of the experiment, the most probable number of organisms originally present in the soil is also calculated.

Information is obtained on the number of organisms and the rate of disappearance of substrate induced by the organisms in culture. Since the media used are selective it is unlikely that this method will allow the development of all organisms involved in the physiological process, so the evidence for activity is only circumstantial and is probably unrelated to activity in the soil.

(b) *The nylon bag method for examining litter breakdown.* Information on the rate of disappearance of leaf litter and the participation of different groups of soil organisms in this process can be obtained by measurement of loss in dry weight of leaves or loss in leaf area of leaf discs placed on the ground surface or buried in soil in nylon-net bags (Gilbert & Bocock, 1962; Edwards & Heath, 1963).

Measurement of loss in dry weight (Gilbert & Bocock, 1962). Leaf litter, collected at leaf fall in nets hanging above the ground, is air-dried to a moisture content of 10–12% (fresh weight). Weighed samples (1·5–2·0 g) of these leaves are placed in nylon nets (untinted hair nets of 1 cm mesh are useful) and are distributed in the area under study where they are anchored in position by plant labels. The nets are removed from the field after six months and the dry weights of the contents are measured after removal of mineral material and visible animals. From these data loss in dry weight can be assessed.

Measurement of loss in leaf area (Edwards & Heath, 1963). 2·5 cm diameter discs of freshly fallen leaves are placed in nylon-mesh bags (10 × 7 cm) of several mesh sizes:

> 7 mm, allowing microbial and soil invertebrate colonization;
>
> 1 mm, allowing microbial and soil invertebrate colonization with the exception of earthworms;
>
> 0·5 mm, allowing microbial, mite, springtail, enchytraeid and small insect colonization;
>
> 0·003 mm, allowing only microbial colonization.

Fifty leaf discs are placed in each bag and buried at 2·5 cm depth in soil. Sampling is carried out at 2-month intervals. The loss in leaf area is estimated by placing a grid of 100 squares of known size over each leaf disc. The number and types of animals found on the discs in each bag can be recorded.

Other techniques have been developed where leaves are individually attached to nylon string, and loss of dry weight of leaves are assayed with time of exposure to the ground surface (Witkamp & Olson, 1963).

The technique, described above, using litter bags of different mesh sizes can be used to show whether initial fragmentation of leaves by soil animals is a necessary prerequisite for extensive microbial decomposition of litter. Witkamp & Crossley (1966) assessed the role of arthropods and microflora in leaf litter decomposition in field experiments using naphthalene (applied at 100 g per m²) to reduce, to 20% of controls, arthropod populations. Comparison of leaf decomposition, indicated by loss in dry weight of leaves in litter bags, was made in naphthalene treated and untreated sites. The data obtained were consistent with the theory that the effects of soil arthropods include the physical fragmenting of leaves and a consequent increase of surface area for microbial attack and leaching.

(c) *Perfusion techniques.* The use of these methods for the isolation of specific groups of soil micro-organisms has been described on p. 55. However, these methods have frequently been used to assess the activity of particular biochemical groups of micro-organisms in a soil sample. Lees & Quastel (1946) describe soil perfusion apparatus, used for studies on nitrification in soil, in which a column of soil is perfused with aerated liquid in a circulatory process (Fig. 7). The liquid contains, in solution, the substance whose metabolism is being studied. The rate of perfusion is adjusted so that waterlogging of the soil column does not occur.

In studies on soil nitrification Lees & Quastel (1946) found a suitable weight of air-dried soil was 20–100 g, and a suitable volume of perfusate, in this case 0·028–0·0071N ammonium solution, was 200–300 ml. The concentration of ammonium solution was adjusted after initial studies on the approximate rate of nitrification.

Lees & Quastel (1946) list the advantages and disadvantages of their method.

Advantages include:

1. The soil column is kept at a constant and evenly distributed water content.

2. Continual liquid flow through the soil column minimizes temperature variations in the soil.

3. The soil column is undisturbed during the experiment and aeration of the soil is effected. Air may be replaced by any required gas mixtures.

4. Sampling of the perfusate is easy, as is the addition of substances (e.g., poisons or microbial inhibitors) to the system.

5. Ionic equilibrium between the soil and solution is quickly attained, but this can be disturbed by metabolic products of the soil.

6. The soil, after perfusion, can be analysed chemically or microbiologically.

7. The apparatus is cheap to make and easily modified for specific purposes.

*Chapter 7*

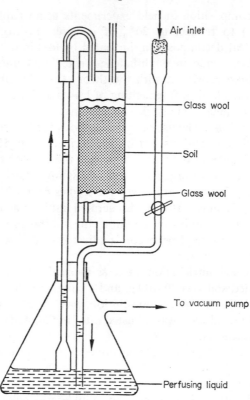

Figure 7. Soil perfusion apparatus, used for studies on nitrification in soil. From Lees H. & Quastel J.H. (1946) *Biochem. J.* **40**, 803.

Disadvantages include:

1. Removal of perfusate for analysis decreases the total volume of perfusate without a corresponding (compensating) decrease in the amount of soil. This error can be minimized by having a large initial volume of perfusate and by removing as small samples as possible for analysis.

2. The apparatus can only be used for studies at a soil moisture content near to waterlogging.

Lees & co-workers (Lees, 1949; Greenwood & Lees, 1956, 1959) have developed apparatus in which oxygen uptake of soil samples can be measured whilst substances under investigation are percolated through the soil sample. Periodically the percolate is tested and assayed chemically.

### 7.3 Measurement of growth rates

At present, there are no methods available to measure accurately the amount of growth in normal soil. The reasons for this are clear; because soil is at a low nutrient level, it is impossible to observe the growth of identifiable colonies for long enough periods of time without amendment or alteration of the soil in various ways. Two approaches have been made to this problem, the first being to study rate of spread of micro-organisms through *sterilized* soil, the second being to study the multiplication of labelled cells placed in the natural environment.

### 7.3.1 Growth rate in sterile soil

(a) *Evans' growth tube* (*Evans, 1955*). Soil is placed in a special tube which consists of a 45 cm length of Pyrex tubing of 2·5 cm diameter to which seven side-arms, 2·5 cm long and 1·25 cm, are fused at 5 cm intervals (see Fig. 8).

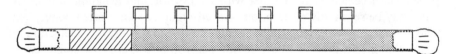

Figure 8. The soil recolonization tube. Key: Hatched, non-sterile inoculum; Stippled, sterile soil. From Evans, E. (1955) *Trans. br. Mycol. Soc.*, **38**, 335.

The soil which has been sieved (4 mm mesh) and adjusted to the required moisture level is added to the tube in 25 g quantities, tamping after each addition to ensure even packing. The completed assembly is wrapped in grease-proof paper and sterilized by steaming for 1 hour on three consecutive days. The tube is inoculated with a substantial amount of non-sterile soil and then weighed. Periodic addition of water through the side arms, starting at the end farthest from the inoculum, is carried out to keep the soil at a constant moisture level.

Soil samples are removed from each side arm at suitable time intervals and plated out to determine whether fungi have grown and reached the sampling point or not. Growth of pure cultures or individual fungi in mixed cultures can be determined. However, sterilization of the soil removes natural competitors and releases additional nutrients, complicating the interpretation of the data.

(b) *Stotzky's replica plating technique* (*Stotzky, 1965*). Soil, sieved or amended as required, is placed in deep Petri dishes and sterilized by autoclaving.

The soil is inoculated centrally with a constant volume loop or dropping pipette. Amendments may also be added in this way. The plates are placed in an incubator containing a pan of water to maintain humidity.

Replications are made from the soil on to agar in Petri dishes at appropriate intervals of time, using the replicator described below. The replicator should be tapped gently so that most adhering soil is removed. Use of stringent aseptic technique will allow replications to be made at least 15–20 times during a 3–4 week period with essentially no contamination.

Growth and distribution rates can be measured in terms of distance of outgrowth from the inoculum or percentage of needles colonized. If mixtures or organisms are present, replication must be made on to selective media to follow the spread of each organism.

A replicator may be made as follows. Holes are drilled in a sheet of acrylic plastic (0·6–0·9 cm thick) about 2·5 cm wider than the Petri dish diameter. Twenty gauge, 3·75 cm flat head stainless steel nails are inserted, another plastic sheet cemented to the first with chloroform and a handle fitted. The nails may be set in any pattern, but spaced radial rows are often convenient. The distance between the nails should be 4–7 mm, allowing detection of small changes in growth rate but not permitting trapping of soil between the needles.

Placing of inocula is often facilitated by the preparation of a template with holes at the required inoculation points.

### 7.3.2   Growth rate in natural environments

(a) *Growth rate of morphologically recognizable organisms in aquatic environments (Brock, 1967)*. This method was described by Brock for the measurement of growth rates of *Leucothrix minor* in water and though it has not been used in soil, because of inherent difficulties of soil as a medium, it is described here because this type of method will be needed for proper assessment of microbial growth in soil. It can only be used with morphologically recognizable organisms growing in transparent media, e.g., water.

The rate of growth of cells in pure culture is determined and the rate of incorporation of tritiated thymidine found by incubating growing cultures in synthetic sea water with tritiated thymidine (1 $\mu$c/ml, 6·7 c/m mole), by preparing autoradiographs at different times, and determining the percentage of radioactive cells (as + or −). From initial experiments of this type it was shown that 1 % of radioactive cells appeared in 0·002 generations, and this can be used as a constant for estimation of growth rate in nature. Thus if it is found that 10% of the cells are radioactive after 30 minutes incubation in natural seawater, it follows that in 30 minutes there would be 0·02 generations. Therefore, the generation time would be $30 \div 0·02 = 1500$ minutes.

(b) *Autoradiography for tracing microbial growth in soil.* If a radioactive substrate, such as plant debris labelled with [14]C, is introduced into soil, the subsequent distribution of the label in the substrate and in the soil microbes utilizing labelled material can be examined by means of autoradiography. Film stripping technique (Grossbard, 1962, 1969). Coverslips are inserted into soil at various distances from the [14]C-labelled material, removed at regular intervals and the microbial growth on them fixed. They are covered with a section of stripping film peeled off an AR10 (Kodak) fine grain autoradiographic stripping film plate and contact is maintained for 2–4 weeks.

After processing, the film is transparent, allowing microscopic examination of the microbial cells under it. If cells are labelled, grains of elemental silver become visible. The distribution and density of these provides an indication of the location and concentration of tracer.

Quantitative measurements of radioactivity may be attempted photochemically using a radiological densitometer.

(c) *Nylon mesh technique.* The application of the nylon mesh technique for the determination of microbial form and arrangement in soil has been described in section 4–6. However, Nagel-de-Boois (1971) has suggested that this method may be modified to allow studies on microbial growth and death rates in soil. To do this pieces of nylon gauze are buried in soil held in a box with a removable glass wall. Thus the nylon gauze can be removed from and replaced in a known area of soil as desired. Therefore, direct observations and measurements of microbial development on known parts of the pieces of gauze can be made regularly over a period of time.

However, the placement of the gauze adjacent to the glass wall of the box means that the soil conditions at that point are highly artificial, i.e., the glass presents a solid barrier to diffusion of gases, liquids and organisms, and the nylon mesh will contain a good deal of water because of condensation on the glass wall.

### 7.3.3 Determination of competitive saprophytic ability

Several methods have been described for confirming the abilities of microbes to colonize a given substrate in competition with other microbes. The term 'competitive saprophytic ability' may be defined as 'the summation of physiological characteristics that make for success in competitive colonization of dead organic substrates' (Garrett, 1956). The concept has been most frequently applied in comparisons of the saprophytic behaviour of root-infecting fungi, but can also be applied to obligate saprophytes.

(a) *Method of Butler (1953) and Garrett (1963)*. A medium of the following composition is prepared:

100 parts sand
3 parts ground maizemeal } by weight
13 parts water

This growth medium is dispensed into flasks, sterilized, and inoculated with an organism whose saprophytic ability is to be tested. Cultures are incubated at 25°C for 4 weeks.

Using sieved soil, a range of soil/inoculum mixtures are prepared, e.g., 100/0, 98/2, 90/10, 50/50, 10/90, 2/98 and 0/100.

A particulate organic substrate, e.g., lengths of sterile wheat straw, suitable for colonization by the test organism is then introduced into the mixtures and distributed as evenly as possible. The systems are incubated for a suitable time, e.g., 4 weeks, maintaining a constant moisture content by regular weighing and water addition.

The ability of the test organism to colonize the substrate in competition with the general soil microflora can be assessed in a number of ways:

1. Agar plate method. Pieces of substrate are removed from the mixtures and surface sterilized (see Section 5.3.2b) to remove surface propagules. After washing with sterile water, they are plated on an appropriate isolation medium and the degree of colonization by the test organism assessed by counting the number of substrate pieces on which it develops from each inoculum/soil mixture.

2. Seedling infection method. Seeds of a plant which can be parasitized by the test organism are inserted into the substrate, e.g., wheat seeds into pieces of wheat straw. After incubation of mixtures, the seeds from each are germinated and the incidence of infection determined.

3. Damp sand method. Pieces of substrate, after surface sterilization and washing, are incubated on a layer of moist, sterile sand in a Petri dish. The presence of sporing fungi on the substrate pieces is detected by direct microscopic examination.

The ability of the test organism to colonize the substrate decreases with decreasing inoculum in the soil mixtures. However, the rate of this decline varies from organism to organism and is inversely related to their competitive saprophytic ability.

(b) *Agar Plate Method* (*Wastie, 1961; Garrett, 1963*). This is a modification of the method just described. In this case, the soil/inoculum mixtures are placed on a plate of nutrient medium which itself serves as the substrate.

After preparation of test organism inoculum/soil mixtures, each mixture is spread evenly in a Petri dish and impregnated with water agar (molten at 45°C). After the agar has solidified, disks of 4 mm diameter are cut out and placed (4 per plate) on the surface of a solidified nutrient medium. After incubation, the number of disks from which the test organism has grown out on to the medium is determined.

## 7.4 Growth of micro-organisms in model systems

### 7.4.1 Introduction

Studies on a known micro-organism or a known group of micro-organisms in pure culture conditions have frequently been carried out in liquid or agar media. However, it is often difficult to relate data obtained from such studies to the behaviour of the test organism in natural soil. Therefore, sterile soil has been used as a medium for studying microbial growth, in which the effects of a variety of environmental conditions on growth can be studied.

Sterile soil media present problems in preparation (i.e., in the choice of sterilization technique) and are difficult to define (chemically and physically) in precise terms. In an attempt to introduce some accurate definition to such media, several types of artificial (but still soil-like) systems have been devised.

Sterile sand is probably the most commonly used growth medium, but materials such as broken procelain, calcium carbonate, sugar, charcoal, vermiculite, aluminum oxide grits, and glass micro-beads have been used in microbiological investigations.

The chemical composition of such systems can be more closely defined than can that of natural soil, this feature being particularly important in studies on plant or microbial nutrition. However, some workers have criticized the application of growth matrices which do not contain colloidal materials, and therefore sand–colloid mixtures have been used in some microbiological studies. Examples of such mixtures are:

1. sand         80%
   pure kaolin   18·5%
   $CaCO_3$      1·0%
   ferric oxide   0·5%

2. sand            97·5%
   bentonite        2·5%
3. sand            97·0%
   bentonite        2·5%
   humus            0·5%

The recent emphasis on a more precise definition of soil physical factors and their effects on microbial growth and activity has led to the development of growth matrices comprising particles of regular size and shape (e.g., glass microbeads of appropriate quality).

As well as the development of systems for experimental studies, which physically resemble soil, there have also been developed systems which chemically resemble soil. These latter include root-like systems for studies on the development of micro-organisms in the root region and continuous culture systems for studies on the growth of microbes at low substrate concentrations.

### 7.4.2  Soil sterilization

Methods of soil sterilization fall into three main categories:
1. Heat (dry or steam)
2. Chemical treatment
3. Gamma irradiation.

All methods cause changes in the treated soil, which must be taken into account in interpreting experiments. The killing of micro-organisms during the sterilization procedure provides a variety of organic substrates which may be utilized by the subsequently introduced test organism(s). Sterilization treatments have also been shown to have effects on the chemical status of the soil. In a comparative study of the effects of different methods of sterilization, Eno & Popenoe (1964) showed that steam sterilization alters the soil more than irradiation which in turn has a greater alteration effect than chemical treatment with methyl bromide.

Steam sterilization causes a significant increase in the amounts of extractable N, P and S, and organic matter value; it also causes a decrease in moisture equivalent percentage.

Gamma irradiation and methyl bromide sterilization both generally increase the extractable N, P and S, but irradiation tends to disrupt soil organic matter more than does the chemical treatment.

Eno & Popenoe (1964) found the effects of sterilization in changing soil properties to be closely associated with soil organic matter content—an organic soil being more altered than a sand.

Steam sterilization: Soil is placed in suitable containers (beakers, Petri dishes) which are covered with paper or foil, and is then sterilized in an autoclave. Eno & Popenoe (1964) used 1 kg soil samples held in covered beakers and sterilized for 6 hours at 20 lb/in.[2]

Gamma irradiation: Soil samples (20–30 g) are placed in polyethylene bags which are closed using elastic bands and are subjected to 3000 kilo-roentgens in a $Co^{60}$ source. At three 4-hour intervals during irradiation the bags of soil are removed, thoroughly mixed and replaced.

Methyl bromide sterilization: Soil samples are placed, in open polyethylene bags, in a vacuum desiccator. Air is evacuated from the desiccator and is replaced with methyl bromide until inside and outside pressures are equalized. The desiccator is closed for 24 hours and then cleared of methyl bromide by flushing several times with sterile air. Ethylene oxide can be used in place of methyl bromide. In the case of gas techniques slight chemical residues may remain after sterilization; these may be toxic to plants.

The details of sterilization technique, i.e., intensity and duration of treatment, will vary from soil to soil and therefore preliminary tests should be performed to ascertain the conditions which ensure proper sterilization of the soil being studied.

### 7.4.3 Other particulate media

(a) *Initial cleaning.* All particulate media require careful cleaning before use in microbiological investigations. In the case of glass microbeads, which have a highly alkaline surface coating when received from the manufacturer, the following procedure has been found efficient.

The microbeads are soaked in 2N nitric acid for 24 hours, the mixture being stirred occasionally. The acid is then decanted and the microbeads washed with many changes, usually more than 40, of tap water until the pH of the water, following decanting, is pH 5. At this point washing with distilled water is begun and is continued until the pH of the water following washing is that of the original distilled water.

The microbeads are left to soak in distilled water for 24 hours, and if there is then no pH change of the water during this period, the water is decanted and the beads dried (50–60°) and stored in stoppered bottles. If an alkaline pH is maintained the beads are returned to acid and the whole procedure repeated.

This general procedure of mineral acid washing followed by thorough water washings is useful for a wide range of particulate inorganic media.

(b) *Sterilization and inoculation with micro-organisms.* The inorganic particulate media commonly used (sand, glass microbeads, grits) are usually steam sterilized by autoclaving. Dry particulate media are placed into suitable containers, e.g., Petri dishes for growth studies, static or aerated systems (see p. 73) for respiration studies. The systems are then steam sterilized (at 130° for 60 minutes).

The sterilized, dry medium is inoculated with the organism under study suspended in a known volume of appropriate nutrient solution. This inoculation is most simply effected by carefully decanting the inoculum down the side of the container into the particulate medium, then placing the inoculated container in a cold room for 24 hours to allow dispersion of the inoculum through the particulate medium before beginning experimentation. More even dispersion of microbial inoculum can be obtained by using containers which allow the particulate medium to be exposed to known suction which in turn allows a known moisture status (pF) to be achieved in the particulate medium.

Bacterial and fungal spore suspensions are easily incorporated into particulate media. However, mycelial suspensions are more difficult to deal with.

(c) *Cleaning of particulate media following experimentation.* Most difficulty is experienced following studies on fungal growth and activity in particulate media because of the development of extensive hyphal networks which frequently bind particles of the growth medium. The following procedure has been useful in cleaning glass microbead systems in which fungi have been grown (Taylor, 1967).

The used microbeads are placed in a beaker with a large volume of water. Rapid swirling of the mixture with a glass rod brings mycelium to the centre of the bead mass surface from whence it can be removed by siphoning. This procedure is repeated until no more mycelium is visible on the surface of the bead mass.

If the beads are very dirty, water can be replaced by 70% sucrose solution. The mixture is stirred and mycelial fragments float to the surface of the liquid. Following thorough washing in this way, the sucrose solution is decanted and the beads are washed with numerous changes of water (Seliwanoff's test is used to detect total removal of sucrose).

After removal of mycelium, the beads are washed in acid and then water as described in Section 7.4.3a.

Efficiency of cleaning of the beads must be tested by assaying for the presence of nitrogen and protein in the washed beads.

### 7.4.4 Artificial roots (collodion membranes, Timonin, 1941)

Glass rods, 0·6 cm diameter and 20 cm long, are heated and drawn to a point. They are then dipped into collodion solution to a depth of 4 cm and then held above the collodion solution until excess solution has dripped off. The rod is rolled between thumb and index finger to allow even coating with collodion. The rod is placed in a support to dry, after which the collodion membrane is removed from the rod and is stored in sterile distilled water.

To determine the permeability of the membrane, it is mounted on the tip of a sterile 10 ml pipette, the pipette and membrane first being filled with sterile nutrient solution, and the membrane is sealed to the pipette with collodion solution. A sterile 1 ml graduate pipette is filled with the same nutrient solution and is fitted to the other end of the 10 ml pipette, and is held in place by means of a rubber stopper. The apparatus is mounted in a vertical position with the collodion membrane suspended in the centre of a pot. The pot is filled with soil, of moisture content between 30 and 60% of the moisture holding capacity. The rate of diffusion of nutrient solution through the membrane per unit time under a constant column is read directly from the 1 ml pipette. Only membranes which allow diffusion not exceeding 0·02 ml solution per hour are used.

These model roots can be used for studies on the rate of development of micro-organisms in artificial rhizospheres.

### 7.4.5 Continuous culture techniques

It is not the intention of this manual to describe methods of carrying out pure culture studies on the physiology of soil organisms in laboratory culture, but the description of artificial systems used to model the physical nature of soil necessitates a few remarks on the limitations of these artificial systems in relation to nutrient supply.

Most artificial matrices have been used in essentially batch culture procedures, i.e., nutrients have been supplied, usually in high concentrations, at the start of the experiment or at intervals throughout the experiment. High concentrations of nutrients are necessary if results are to be obtained, otherwise they become exhausted shortly after the start of the experiment. During the metabolism of these nutrients, profound chemical changes take place so that the environment of the organisms changes during the progress of the experiment and high concentrations of other metabolites may appear. It is difficult, therefore, to interpret the significance of the results.

In the natural environment nutrients are present in low concentrations and are being constantly added and removed. The rate of addition and disappearance is often governed by the environment. It is possible to model these aspects of the natural environment by use of a chemostat (Brock, 1966). In the chemostat, the inflowing nutrient supply has an excess of all nutrients, except one which is limiting. Growth rate is thus determined by the nutrient flow rate and as the flow rate is decreased, the growth rate decreases until a point is reached at which the population stops growing. The effect of parameters such as temperature, pH, etc., on these growth characteristics can be easily determined whilst the conditions within the culture vessel remain constant. Since low nutrient concentrations may be used, one can consider that, at least in this respect, the cultural conditions are ecologically more realistic than those obtained in batch culture.

## 7.5   Inferential assessment of activity

All the methods described above involve measurements of microbial metabolic activities or growth as they occur in field or laboratory conditions. It is possible to obtain more indirect evidence of levels of activity from measurements of the amount of viable cells present in a soil at any one time. This provides an indication of previous activity levels as inferred from the quantity of cell material produced. Such information can be particularly useful when making comparisons between different soils or soil treatments in field conditions, where assessment of metabolic activities and growth as they occur is difficult (see above).

Any technique which provides information on the quantity of microbes in soil can be applied. It may be sufficient to simply compare frequency or numbers of microbes in different soil systems, but for more accurate comparison, the biomass produced by previous activity should be calculated. Some examples of available techniques are given below, together with an assessment of their applicability.

(i) Plating of washed soil materials (see p. 43). These methods provide a crude assessment of the frequency of fungi existing mainly as hyphae on soil substrates, i.e., percentage of plated particles colonized.

(ii) Dilution plates (see p. 64). These give an estimate of the number of bacterial cells, actinomycete and fungal spores in a soil, expressed as number or biomass per unit of soil.

(iii) Jones & Mollison slides (see p. 61). From these lengths of fungal hyphae and numbers of bacterial cells per unit of soil are obtained which can be expressed as biomass.

(iv) Fluorescent antibody technique (see p. 22). An estimate of numbers of cells of a known organism on surfaces in soil can be obtained.

None of these techniques gives information on the time taken for biomass to be produced.

H

# 8

# Identification of Soil Organisms

## 8.1 Introduction

The quantity of published taxonomic work on soil micro-organisms is encyclopaedic and to reproduce it here, even in condensed form, is clearly impossible. The following bibliography is not exhaustive, but gives details of books most frequently used in laboratories in which studies on soil organisms are in progress. In addition to identification keys, reference is made to monographs on individual groups or genera, general texts which include information on techniques and diagnostic procedures, and some books or papers containing taxonomic theory.

## 8.2 General taxonomic theory

AINSWORTH G.C. & SNEATH P.H.A. (1962). Microbial classification. *Symp. Soc. gen. Microbiol.* **12**. Cambridge: Cambridge University Press.
SOKAL R.R. & SNEATH P.H.A. (1963). *Principles of numerical taxonomy.* San Francisco & London: Freeman.

## 8.3 Fungi

### 8.3.1 General information

AINSWORTH G.C. & BISBY G.R. (1963). *Dictionary of the fungi.* 5th ed. Kew, Surrey: Commonwealth Mycological Institute.
CLEMENTS F.E. & SHEAR C.L. (1931). *The genera of fungi.* New York: H.W. Wilson & Co.
COMMONWEALTH MYCOLOGICAL INSTITUTE, KEW (1960). Herb. I.M.I. Handbook: methods in use at the Commonwealth Mycological Institute.

### 8.3.2 Keys and monographs for identification

BARNETT H.L. (1960). *Illustrated genera of imperfect fungi.* Minneapolis: Burgess Publishing Co.

BARRON G.L. (1968). *The genera of Hyphomycetes from soil.* Baltimore: Williams and Wilkins.

DOMSCH K. & GAMS W. (1970). *Pilze aus Agrarboden.* Fischer, Stuttgart.

GILMAN J.C. (1957). *Manual of soil fungi.* Ames, Iowa: Iowa State University Press.

GROVE W.B. (1937). *British stem and leaf fungi*, vols. 1 & 2. Cambridge: University Press.

LINNEMAN G. (1941). *Die Mucorineen—Gattung* Mortierella *Coemans.* Pflanzenforschung 23 : 1–64.

RAPER K.B. & FENNELL D.I. (1965). *The genus* Aspergillus. Baltimore: Williams and Wilkins.

RAPER K.B. & THOM C. (1949). *A manual of the penicillia.* Baltimore: Williams and Wilkins.

ZYCHA H. (1935). *Mucorineae.* Lehre, Germany: J. Cramer.

## 8.4 Bacteria and actinomycetes

### 8.4.1 Keys for identification

BREED R.S., MURRAY E.G.D. & SMITH N.R. (1957). *Bergey's manual of determinative bacteriology.* 8th ed. London: Baillière, Tindall & Cox.

BUCHANAN R.E., HOLT J.G. & LESSEL E.F. (1966). *Index Bergeyana.* Edinburgh, London : E. & S. Livingstone.

COWAN S.T. & STEEL K.J. (1961). Diagnostic tables for the common medical bacteria. *J. Hyg. Camb.* **59**, 357.

COWAN S.T. & STEEL K.J. (1965). *Manual for the identification of medical bacteria.* Cambridge, Cambridge University Press.

HÜTTER R. (1967). *Systematik der Streptomyceten.* Basel: Kroger, A.G.

SKERMAN V.B.D. (1967). *A guide to the identification of the genera of bacteria.* 2nd ed. Baltimore: Williams and Wilkins.

SMITH N.R., GORDON R.E. & CLARK F.E. (1952). *Aerobic spore-forming bacteria.* U.S Dept. Agric. Monograph 16, Washington, D.C.

WILLIAMS S.T., DAVIES F.L. & CROSS T. (1968). *Identification of genera of the Actinomycetales.* In *Identification methods for microbiologists B.* Ed. Gibbs, B.M. & Shapton, D.A. London, New York: Academic Press.

### 8.4.2 General methods and techniques (and books also including keys)

GIBBS B.M. & SKINNER F.A. (1966) *Identification methods for microbiologists A.* London, New York: Academic Press (includes treatments on *Pseudo-*

*monas, Xanthomonas, Chromobacterium,* Enterobacteriaceae, *Staphylococcus* and *Micrococcus,* lactic acid bacteria, *Clostridium, Mycobacterium, Streptomyces* and rapid methods for testing bacteria).

GIBBS B.M. & SHAPTON D.A. (1968). *Identification methods for microbiologists B.* London & New York: Academic Press (includes treatments of acetic acid bacteria, *Azotobacter, Acinetobacter, Rhizobium,* yellow pigmented rods, *Aeromonas* and *Vibrio, Bacillus,* the Actinomycetales and yeasts).

GORDON R.E. (1968). The taxonomy of soil bacteria. In *The Ecology of Soil Bacteria.* Ed. Gray T.R.G. and Parkinson D. Liverpool: Liverpool University Press (Bibliography and methods for dealing with some *Nocardia* and *Streptomyces* strains). See also other papers in the same volume on matters relating to taxonomy of soil bacteria.

MARTIN S.M. (1962). *Culture collections: perspectives and problems.* Ottawa: University of Toronto Press.

SHIRLING E.B. & GOTTLIEB D. (1966). Methods for characterization of *Streptomyces* species. *Int. J. System. Bact.* **16,** 313–340.

WAKSMAN S.A. (1967). *The actinomycetes, a summary of current knowledge.* New York: Ronald Press.

# 9

# Media for Isolation of Soil Micro-organisms

## 9.1 Introduction

A great variety of media have been used for the isolation of soil micro-organisms; almost every worker has his own favourite ones and the list given here is only intended to give some idea of the range available.

## 9.2 Solidifying agents

Media may be prepared in broth form, or solidified in various ways for plating. The most commonly used solidifying agents are agar and silica gel.

Agar is available as a commercial preparation. In recent years, agar quality has improved enormously and although at one time it was necessary to use concentrations of 2% to solidify media, many modern forms can be used at much lower concentrations; makers usually give the necessary information with the product. Consequently, the concentration of agar in the various media have not been given. It should also be noted that some agars contain many impurities and this may cause precipitation in the media. In these cases, it may be necessary to filter the medium, prior to sterilization.

Silica gel has a number of advantages over agar. It is inorganic and wholly inert and cannot act as a substrate for growth. It is used mainly to solidify media on which are grown organisms sensitive to organic matter, or where precisely defined conditions are essential. Silica gels may be made using the procedure outlined by Funk & Krulwich (1964).

Two solutions are prepared:

(1) 20% aqueous o-phosphoric acid (sterile)

(2) 10 g powdered silica gel (grade 923, 100–200 mesh) or silicic acid (reagent grade) dissolved in 100 ml 7% (w/v) aqueous KOH by heating to form potassium silicate.

Freshly made solution (2) is dispensed in 20 ml portions and autoclaved at 15 lb/in.$^2$ for 15 minutes.

Equal volumes of the sterilized solution (2) and sterile double strength fluid growth medium are mixed, rapidly adding about 4·0 ml of the sterile o-phosphoric acid solution (1). The mixed solutions are immediately poured into sterile dishes. Gelation begins one minute after addition of that amount of acid which brings the reaction of the mixture to pH 7·0. The gel becomes firm in about 15 minutes and undergoes syneresis. The water of syneresis should be decanted or evaporated off. Plates should be incubated in a moist atmosphere to prevent drying and cracking of the medium.

### 9.3   Selective agents

Media may be made partially selective for fungi, bacteria or actinomycete by alterations to the reaction of the medium or by addition of selective agents. pH values of 5·0–5·5 are selective for fungi while neutral and slightly alkaline pH values are selective for bacteria and actinomycetes. In certain extreme environments, acidophilic bacteria and actinomycetes may occur and, likewise, fungi growing in alkaline conditions can be found.

Micro-organisms may also be selected for by the use of inhibitory additives. For the reduction of fungal development on isolation media, the addition of actidione (50 $\mu$g/ml) and nystatin (50 $\mu$g/ml) is suitable. For the reduction of most bacterial growth the addition of aureomycin (30 $\mu$g/ml) is recommended while the suppression of Gram-positive bacteria only, can be achieved by adding penicillin (1·0 $\mu$g/ml). Conversely, for the selective suppression of Gram-negative bacteria, polymixin B (5·0 $\mu$g/ml) can be incorporated in the medium.

### 9.4   Soil extract media

Soil extracts often form the basis of media of relatively low selectivity. They are suitable for fungi, bacteria or actinomycetes, providing one or more of the above selective agents are added. James (1958) has suggested the following procedure for making extracts.

One kg of soil is autoclaved with 1 litre water for 20 minutes at 20 lb/in.[2]. The liquid is strained and restored to 1 litre in volume. If it is cloudy, a little calcium sulphate is added and after being allowed to stand, it is filtered. The extract may be sterilized and solidified with agar as it is, or after the addition of other nutrients, e.g., 0·025% $K_2HPO_4$ or 0·1% glucose, 0·5% yeast extract and 0·02% $K_2HPO_4$.

## 9.5 Other general isolation media

### 9.5.1 Media for the isolation of bacteria (heterotrophs)

(a) *Peptone yeast extract agar* (Goodfellow, *et al.*, 1968). Peptone, 5·0 g; yeast extract (spray dried), 1·0 g; ferric phosphate, 0·01 g; agar; distilled water, 1 litre, pH 7·2.

(b) *Bunt & Rovira's medium* (Bunt & Rovira, 1955). $K_2HPO_4$, 0·4 g; $(NH_4)_2HPO_4$, 0·5 g; $MgSO_4.7H_2O$, 0·05 g; $MgCl_2$, 0·1 g; $FeCl_3$, 0·01 g; $CaCl_2$, 0·1 g; peptone 1·0 g; yeast extract, 1·0 g; soil extract 250 ml; distilled water, 750 ml; agar; pH 7·4. Steam for 30 minutes. Filter and autoclave at 10 lb/in.$^2$ for 20 minutes.

(c) *Nutrient agar*. Lab-lemco, 1·0 g; yeast extract, 2·0 g; peptone, 5·0 g; sodium chloride, 5·0 g; agar; distilled water, 1 litre; pH 7·3.

(d) *Thornton's medium* (Thornton, 1922). $K_2HPO_4$, 1·0 g; $MgSO_4.7H_2O$, 0·2 g; $CaCl_2$, 0·1 g; NaCl, 0·1 g; $FeCl_3$, trace; $KNO_3$, 0·5 g; asparagine, 0·5 g; mannitol, 1·0 g; agar; distilled water, 1 litre.

### 9.5.2 Media for the isolation of actinomycetes

(a) *Starch casein agar* (Küster & Williams, 1966). Starch, 10·0 g; casein (vitamin free), 0·3 g; $KNO_3$, 2·0 g; NaCl, 2·0 g; $K_2HPO_4$, 2·0 g; $MgSO_4.7H_2O$, 0·05 g; $CaCO_3$, 0·02 g; $FeSO_4.7H_2O$, 0·01 g; agar; distilled water, 1 litre; pH 7·2. Can be improved by addition of antifungal antibiotics.

(b) *Lingappa & Lockwood's medium* (Lingappa & Lockwood, 1962). Chitin (ball milled or reprecipitated), 2·0 g; agar; distilled water, 1 litre. The following salts may be added if desired: $K_2HPO_4$, 0·7 g; $KH_2PO_4$, 0·3 g; $MgSO_4.5H_2O$, 0·5 g; $FeSO_4.7H_2O$, 0·01 g; $ZnSO_4$, 0·001 g; pH 7·0.

### 9.5.3 Media for the isolation of fungi

(a) *Czapek-Dox agar*. Sucrose, 30·0 g; $NaNO_3$, 2·0 g; $K_2HPO_4$, 0·1 g; KCl, 0·5 g; $MgSO_4.7H_2O$, 0·5 g; $FeSO_4$, trace; agar; distilled water, 1 litre. pH 5·5. Dissolve separately and add $FeSO_4$ last. If desired 6·5 g of yeast extract may be added.

(b) *Potato dextrose agar*. Potatoes, peeled and sliced, 200·0 g; dextrose, 20 g; agar; distilled water, 1 litre. Cook the potatoes for 1 hour in a steamer in 500 ml water. Strain the resulting juice and pour into the melted agar medium. Adjust the volume with water and then add dextrose.

(c) *Malt extract agar*. Malt extract, 20·0 g; dextrose, 20·0 g; peptone,

1·0 g; 1 litre water; the concentration of malt extract may be lowered if desired.

(d) *Oatmeal agar*. Rolled oats, 65·0 g; agar; water, 1 litre; Cook the rolled oats in a double boiler, or heat to 70°C for 1 hour. Filter and add agar to filtrate and sterilize at 10 lb for 30 minutes.

## 9.6 Media for the isolation of organisms of particular physiological groups
### 9.6.1 Micro-organisms involved in carbon compound transformations

(a) *Cellulose agar* (Eggins & Pugh, 1961). $NaNO_3$, 0·5 g; $K_2HPO_4$, 1·0 g; $MgSO_4.7H_2O$, 0·5 g; $FeSO_4.7H_2O$, 0·01 g; cellulose (ball milled), 12·0 g; agar; distilled water, 1 litre.

(b) *Chitin agar*. Ball milled, purified chitin, 10·0 g; $MgSO_4.7H_2O$, 1·0 g; $K_2HPO_4$, 1·0 g; distilled water, 1 litre.

(c) *Starch agar*. 0·2% soluble starch may be added to any suitable growth medium as an alternative or additional carbohydrate. Starch hydrolysis is shown by flooding incubated plates with an iodine solution and noting clear zones.

(d) *Pectate gels* (Paton, 1959). Two solutions are prepared:

(i) Calcium agar; peptone, 0·5%; Lab-lemco, 0·5%; calcium lactate, 0·5%, agar; distilled water. Pour into Petri dishes.

(ii) Pectin solution: mix 2 g pectate powder with absolute alcohol to form a thin paste. Add water to give 100 ml and add 0·1 g of the disodium salt of ethylene diamine tetra acetic acid (EDTA). Sterilize and pour this on the agar. A gel forms in 12–24 hours at 27°

(e) *Photosynthetic non-sulphur bacteria* (van Niel, 1944). $(NH_4)_2SO_4$, 1·0 g; $K_2HPO_4$, 0·5 g; $MgSO_4.7H_2O$, 0·2 g; $NaHCO_3$, 5·0 g; $NaCl$, 2·0 g; sodium formate or succinate, 1·5 g; distilled water, 1 litre; pH 7·0. Adjusted with $H_3PO_4$.

### 9.6.2 Micro-organisms involved in nitrogen transformations

(a) *Free living nitrogen-fixing bacteria*. $K_2HPO_4$, 0·5 g; $MgSO_4.7H_2O$, 0·2 g; $NaCl$, 0·2 g; $MnSO_4.4H_2O$, trace; $FeCl_3.6H_2O$, trace; agar; distilled water. To each 50 ml add 1 ml of 10% mannitol solution, sterilized by filtration.

(b) *Symbiotic nitrogen fixing bacteria*-Rhizobium

(i) Yeast extract, mannitol medium (Allen, 1957). Mannitol, 10·0 g; $K_2HPO_4$, 0·5 g; $MgSO_4.7H_2O$, 0·2 g; $NaCl$, 0·1 g; $CaCO_3$, 3·0 g; yeast extract (10%), 100 ml; agar.

(ii) Ham's medium (pers. comm.). Mannitol, 10·0 g; $K_2HPO_4.3H_2O$,

0·65 g; $MgSO_4.7H_2O$, 0·2 g; NaCl, 0·1 g; yeast extract (spray dried), 1·0 g; agar; water, 1 litre.

(c) *Dentrifying bacteria* (de Barjac, 1952). Media should contain nitrate (0·2%) as a sole source of nitrogen. This is not a selective medium for organisms reducing nitrate to nitrogen gas, but it is suitable for organisms that reduce nitrate in some way, e.g., $KNO_3$, 2·0 g; glucose, 10·0 g; $CaCO_3$, 5·0 g; Winogradsky's mineral salt solution, 1 litre ($K_2HPO_4$, 5·0 g; $MgSO_4.7H_2O$, 2·5 g; NaCl, 2·5 g; $FeSO_4.7H_2O$, 0·05 g; $MnSO_4$, 0·05 g; distilled water, 1 litre).

(d) *Ammonifying bacteria.* Any medium containing single amino-acids (e.g., tyrosine, asparagine, 0·05%), proteins (3%), or casein hydrolysate may be used.

(e) *Nitrifying bacteria* (Lewis & Pramer, 1958). $Na_2HPO_4$, 13·5 g; $KH_2PO_4$, 0·7 g; $MgSO_4.7H_2O$, 0·1 g; $NaHCO_3$, 0·5 g; $(NH_4)_2SO_4$, 2·5 g; $FeCl_3.6H_2O$, 14·4 mg; $CaCl_2.7H_2O$, 18·4 mg; water, 1 litre; pH 8·0. Distribute in 100 ml aliquots. Inoculate with enriched soil. After 8 days at 28°, the ratio of nitrifiers/heterotrophs is about 250/1. Dilutions of $10^7–10^9$ can give cultures 90% free of contamination. See Lewis & Pramer (1958) for further details.

### 9.6.3 Micro-organisms involved in sulphur transformations

(a) *Photosynthetic sulphur bacteria* (van Niel, 1931). $NH_4Cl$, 1·0 g; $K_2HPO_4$, 0·5 g; $MgCl_2$, 0·2 g; $NaHCO_3$, 0·5 g; $Na_2S$, 1·0 g; water, 1 litre, pH 8·5 for purple sulphur bacteria, pH 7·0 for green sulphur bacteria.

(b) *Thiobacillus thiooxidans.* $(NH_4)_2SO_4$, 0·4 g; $KH_2PO_4$, 4·0 g; $MgSO_4.7H_2O$, 0·5 g; $CaCl_2$, 0·25 g; $FeSO_4$, 0·01 g; powdered sulphur 10·0 g on $Na_2S_2O_7$, 5·0 g; water, 1 litre. pH 7·0.

(c) *Thiobacillus denitrificans.* $KNO_3$, 1·0 g; $Na_2HPO_4$, 0·1 g; $Na_2S_2O_7$, 2·0 g; $NaHCO_3$, 0·1 g; $MgCl_2$, 0·1 g; water, 1 litre. pH 7·0.

(d) *Sulphate reduction* (Adams & Postgate, 1961). For enrichment culture $K_2HPO_4$, 0·5 g; $NH_4Cl$, 1·0 g; $Na_2SO_4$, 1·0 g; $CaCl_2.6H_2O$, 0·1 g; $MgSO_4.7H_2O$, 2·0 g; sodium lactate (70% solution), 3·5 g; yeast extract (spray dried), 1·0 g; $FeSO_4.7H_2O$, 0·002 g; water, 1 litre. pH 7·5. Autoclave 20 minutes/15 lb/in.[2]. Filter and resterilize. For heterotrophic growth of sulphate reducers, add 0·01% (w/v) sodium thioglycollate and 0·05% (w/v) ferrous ammonium sulphate pH 7·4 ± 0·3 and put in stoppered bottles.

# References

ADAMS M.E. & POSTGATE J.R. (1960). On sporulation in sulphate-reducing bacteria. *J. gen. Microbiol.* **24,** 291–294.

ALEXANDER F.E.S. & JACKSON R.M. (1955). Preparations of sections for study of soil micro-organisms. In *Soil Zoology,* ed. D.K.McE. Kevan. pp. 443–441. Butterworths, London.

ALEXANDER M. (1961). *Introductions to soil microbiology.* Wiley, New York.

ALLEN, O.N. (1957). *Experiments in soil bacteriology,* 3rd edn. Burgess, Minneapolis.

ANTHONY E.H. (1970). Bacteria in estuarine (Bras d'Or Lake) sediment. *Can. J. Microbiol.* **16,** 373–389.

ARISTOVSKAYA T.V. & PARINKINA O.M. (1961). New methods of studying soil micro-organism associations. *Soviet Soil Sci.,* 12–20.

BABIUK L.A. & PAUL E.A. (1970). The use of fluoroscein isothiocyanate in the determination of the bacterial biomass of grassland soil. *Can. J. Microbiol.* **16,** 57–62.

BAILEY N.T.J. (1959). *Statistical methods in biology.* English University Press, London.

DE BARJAC H. (1952). La puissance denitrifiante du sol. Mise au point d'une technique d'evaluation. *Annls Inst. Pasteur, Paris,* **83,** 207–212.

BAVER L.D. (1948). *Soil physics.* 2nd edn. John Wiley, New York.

BAXBY P. & GRAY T.R.G. (1968). Chitin decomposition in soil. I. Media for isolation of chitinoclastic micro-organisms from soil. *Trans. br. mycol. Soc.* **51,** 287–292.

BLACK C.A. *et al.* (1965). *Methods of soil analysis. II. Chemical and microbiological properties.* American Society of Agronomy, Madison.

BLAIR I.D. (1945). Techniques for soil fungus studies. *N.Z. J. Sci. Tech.* 258–271.

BOHLOOL B.B. & SCHMIDT E.L. (1968). Non-specific staining: its control in immunofluorescence examination of soil. *Science, N.Y.* **162,** 1012–1014.

BRIERLEY W.B., JEWSON S.T. & BRIERLEY M. (1928). The quantitative study of soil fungi. *1st int. Congr. Soil Sci.* **3,** 48–71.

BROCK T.D. (1966). *Principles of microbial ecology.* Prentice Hall, New Jersey.

BROCK T.D. (1967). Bacterial growth rate in the sea: direct analysis by thymine autoradiography. *Science, N.Y.* **155,** 81–83.

BROWN J.C. (1958). Soil fungi of some British sand dunes in relation to soil type and succession. *J. Ecol.* **46,** 641–664.

BUDDENHAGEN I.W. (1965). The ecology of plant pathogenic bacteria in soil. In *Ecology of soil-borne plant pathogens,* ed. K.F. Baker & W.C.Snyder. pp. 269–284. University of California Press, Berkeley.

BUNT J.S. & ROVIRA A.D. (1955). Microbiological studies of some subantarctic soils. *J. Soil Sci.* **6,** 119–128.

BURGES A. & NICHOLAS D.P. (1961). Use of soil sections in studying amounts of fungal hyphae in soil. *Soil Sci.* **92,** 25–29.

BUTLER F.C. (1953). Saprophytic behaviour of some cereal root rot fungi. I. Saprophytic colonization of wheat straw. *Ann. appl. Biol.* **40,** 284–297.

CARTER H.P. & LOCKWOOD J.L. (1957). Methods for estimating numbers of soil micro-organisms lytic to fungi. *Phytopathology* **47**, 151–154.

CASIDA L.E. JR. (1962). On the isolation and growth of individual microbial cells from soil. *Can. J. Microbiol.* **8**, 115–119.

CASSIE R.M. (1962). Frequency distribution models in the ecology of plankton and other organisms. *J. Anim. Ecol.* **31**, 65–92.

CHESTERS C.G.C. (1940). A method for isolating soil fungi. *Trans. br. mycol. Soc.* **24**, 352–355.

CLARK F.E. (1965). Agar plate method for total microbial count. In *Methods of soil analysis. II*, ed. C.A.Black *et al.*, pp. 1460–1466. American Society of Agronomy, Madison.

CLARK F.E., JACKSON R.D. & GARDNER H.R. (1962). Measurement of microbial thermo-genesis in soil. *Proc. Soil Sci. Soc. Am.* **26**, 155–160.

CLARKE J.H. (1961). *Studies on fungi associated with roots of certain species of Allium.* Ph.D. Thesis, University of Liverpool.

CLARKE J.H. & PARKINSON D. (1960). A comparison of three methods for the assessment of fungal colonization of seedling roots of leek and broad bean. *Nature, Lond.* **188**, 166–167.

CLINE M.G. (1944). Principles of soil sampling. *Soil Sci.* **58**, 275–288.

CONWAY E.J. (1950). *Microdiffusion and volumetric error.* Lockwood, London.

COOKE R.C. (1962). Behaviour of nematode-trapping fungi during decomposition of organic matter in soil. *Trans. br. mycol. Soc.* **45**, 314–320.

CRIPPS R.E. & NORRIS J.R. (1969). A soil perfusion apparatus. *J. appl. Bact.* **32**, 259–260

DOMSCH K. & GAMS W. (1970). *Pilze aus Agrarboden.* Fischer, Stuttgart.

DUBEY H.D. (1957). A method for soil moisture control in nitrification studies. *Soil Biol. Newsletter* **7**, 35.

DUDDINGTON C. (1954). Nematode destroying fungi in agricultural soil. *Nature, Lond.* **173**, 500–501.

DUDDINGTON C. (1955). Notes on the technique of handling predacious fungi. *Trans. br. mycol Soc.* **38**, 97–103.

DUNNING W.J. (1955). The application of mass spectroscopy to problems of chemical analysis. *Quart. Rev.* **9**, 23–50.

DURBIN R.D. (1965). A compilation of solutions for maintaining constant relative humidities. *Pl. Dis. Reptr* **49** (ii), 948–54.

EDWARDS C.A. & HEATH G.W. (1963). The role of soil animals in breakdown of leaf litter. In *Soil organisms*, ed. J.Doeksen & J.van der Drift. pp. 76–84. North Holland Publishing Co., Amsterdam.

EGDELL J.W., CUTHBERT W.A., SCARLETT C.A., THOMAS S.B. & WESTMACOTT J. (1960). Some studies of the colony count technique for soil bacteria. *J. appl. Bact.* **23**, 69–80.

EGGINS H.O.W. & PUGH G.J.F. (1961). Isolation of cellulose-decomposing fungi from soil. *Nature, Lond.* **193**, 94–95.

ENO C.F. & POPENOE H. (1964). Gamma radiation compared with steam and methyl bromide as a soil sterilizing agent. *Proc. Soil Sci. Soc. Am.* **28**, 533–535.

EVANS E. (1955). Survival and recolonization by fungi in soil treated with formalin or carbon disulphide. *Trans. br. mycol. Soc.* **38**, 335–346.

FÅHRAEUS G. (1957). The infection of clover root hairs by nodule bacteria studied by a simple glass slide technique. *J. gen. Microbiol.* **16**, 374–381.

FISHER R.A. & YATES F. (1963). *Statistical tables for biological, agricultural and medical research.* 6th ed. Oliver and Boyd, Edinburgh.

FUNK H.B. & KRULWICH T.A. (1964). Preparation of clear silica gels that can be streaked. *J. Bact.* **88**, 1200–1201.

GAMS W. & DOMSCH K. (1967). Beitrage zur Anwendung der Bodenwaschtechnik für die Isolierung van Bodenpilzen. *Arch. Mikrobiol.* **58**, 134–144.

GARRETT S.D. (1956). *Biology of root infecting fungi.* Cambridge University Press, Cambridge.

GARRETT S.D. (1963). *Soil fungi and soil fertility.* Pergamon, Oxford.

GILBERT O.J. & BOCOCK K.L. (1962). Some methods of studying the disappearance and decomposition of leaf litter. In *Progress in Soil Zoology*, ed. P.W.Murphy. pp. 348–352. Butterworths, London.

GRAHAM R.K. & CAIGER P. (1969). Fluorescence staining for the determination of cell viability. *Appl. Microbiol.* **17**, 489–490.

GRAY T.R.G. (1967). Stereoscan electron microscopy of soil micro-organisms. *Science, N.Y.* **155**, 1668–1670.

GRAY T.R.G. & BELL T.F. (1963). The decomposition of chitin in an acid soil. In *Soil organisms*, ed. J.Doeksen & J.van der Drift. pp. 222–230. North Holland Publishing Co., Amsterdam.

GRAY T.R.G. & LOWE W.E. (1967). Techniques for studying cutin decomposition in soil. *Bact. Proc.* 3.

GRAY T.R.G. & WILLIAMS S.T. (1971). *Soil micro-organisms.* Oliver & Boyd, Edinburgh.

GREENWOOD D.J. & LEES H. (1956). Studies on the decomposition of amino-acids in soil. I. A preliminary survey of techniques. *Pl. Soil* **7**, 253–268.

GREENWOOD D.J. & LEES H. (1959). An electrolytic rocking percolator. *Pl. Soil* **11**, 87–92.

GROSSBARD E. (1962). Autoradiography of bacteria and Streptomycetaceae by the stripping film technique. *Nature, Lond.* **193**, 853.

GROSSBARD E. (1969). A visual record of the decomposition of $^{14}$C-labelled fragments of grasses and rye added to soil. *J. Soil Sci.* **20**, 38–51.

HARLEY J.L. & WAID J.S. (1955). A method of studying active mycelia on living roots and other surfaces in the soil. *Trans. br. mycol. Soc.* **38**, 104–118.

HERING T.F. (1966). An automatic soil-washing apparatus for fungal isolation. *Pl. Soil* **25**, 195–200.

HILL I.R. & GRAY T.R.G. (1967). Application of the fluorescent antibody technique to an ecological study of bacteria in soil. *J. Bact.* **93**, 1888–1896.

HOWARD P.J.A. (1968). The use of Dixon and Gilson respirometers in soil and litter respiration studies. *Merlewood Research and Development Paper, Grange, Lancs, U.K.* **5**.

JAMES N. (1958). Soil extract in soil microbiology. *Can. J. Microbiol.* **4**, 363–370.

JAMES N. & SUTHERLAND M.L. (1939). The accuracy of the plating method for estimating the numbers of soil bacteria, actinomycetes and fungi in the dilution plated. *Can. J. Res. C* **17**, 72–86.

JENKINSON D.S. (1966). Studies on the decomposition of plant material in soil. II. Partial sterilization and the soil biomass. *J. Soil Sci.* **17**, 280–302.

JENNY H. & GROSSENBACHER K. (1963). Root-soil boundary zones as seen by the electron microscope. *Proc. Soil Sci. Soc. Am.* **27**, 273–277.

JENSEN V. (1962). Studies on the microflora of Danish beech forest soils. I. The dilution plate count technique for the enumeration of bacteria and fungi in soil. *Zentbl. Bakt. ParasitKde., Abt. II*, **116**, 13–32.

JENSEN V. (1968). The plate count technique. In *The ecology of soil bacteria*, ed. T.R.G.Gray & D.Parkinson. pp. 158–170. Liverpool University Press, Liverpool.

JONES P.C.T. & MOLLISON J.E. (1948). A technique for the quantitative estimation of soil micro-organisms. *J. gen. Microbiol.* **2**, 54–69.

KATZNELSON H. & ROUATT J.W. (1957). Manometric studies with rhizosphere and non-rhizosphere soil. *Can. J. Microbiol.* **3**, 673–678.

KELNER A. (1948). A method for investigating large microbial populations for antibiotic activity. *J. Bact.* **56**, 1957–1962.

KEULEMANS A.I.M. (1959). *Gas Chromatography.* 2nd edn. Reinhold, New York.

KIBBLE R.A. (1966). *Physiological activity in a pinewood soil.* Ph.D. thesis, University of Liverpool.

KRASSILNIKOV N.A. (1961). *Soil micro-organisms and higher plants.* Israel Program for Scientific Translations, Jerusalem.

KUBIENA W.L. (1938). *Micropedology.* Collegiate Press, Ames, Iowa.

KÜSTER E. & WILLIAMS S.T. (1964). Selection of media for isolation of streptomycetes. *Nature, Lond.* **202**, 928–929.

LEDEBERG J. & LEDEBERG E.M. (1952). Replica plating and indirect selection of bacterial mutants. *J. Bact.* **63**, 399–406.

LEDINGHAM R.J. & CHINN S.H.F. (1955). A flotation method for obtaining spores of *Helminthosporium sativum* from soil. *Can. J. Bot.* **38**, 298–303.

LEES H. (1949). The soil percolation technique. *Pl. Soil* **1**, 221–239.

LEES H. & QUASTEL J.H. (1946). Biochemistry of nitrification in soil. I. Kinetics of, and the effect of poisons on, soil nitrification as studied by a soil perfusion technique. *Biochem. J.* **40**, 803–815.

LENHARD G. (1956). Die Dehydrogenaseaktivitat des Bodens als Mass für Microorganismentatigkeit in Boden. *Z. PflErnähr. Düng. Bodenk.* **73**, 1–11.

LEWIS R.F. & PRAMER D. (1958). Isolation of *Nitrosomonas* in pure culture. *J. Bact.* **76**, 524–528.

LINGAPPA B.T. & LOCKWOOD J.L. (1963). Direct assay of soils for fungistasis. *Phytopathology* **53**, 529–531.

LINGAPPA Y. & LOCKWOOD J.L. (1962). Chitin media for selective isolation and culture of actinomycetes. *Phytopathology* **52**, 317–323.

LOCCI R. & QUARONI S. (1969). A simple freezing technique for microscopical examination of soil. *Rivista di Patologia Vegetale (IV)* **5**, 213–216.

LOWRY O.H., ROSEBROUGH N.J., FARR A.L. & RANDALL R.J. (1951). Protein measurement with the Folin phenol reagent. *J. biol. Chem.* **193**, 265–275.

McGARITY J.W. & MYERS M.G. (1967). A survey of urease activity in soils of Northern New South Wales. *Pl. Soil* **27**, 217–238.

McLENNAN E. (1928). The growth of fungi in the soil. *Ann. appl. Biol.* **15**, 95–109.

METSON A.J. (1956). *Chemical analysis for soil survey samples.* New Zealand DSIR, Wellington.

MEYNELL G.C. & MEYNELL E.W. (1965). *Theory and practice in experimental bacteriology.* Cambridge University Press, Cambridge.

MILLAR W.N. & CASIDA L.E. JR. (1970). Evidence for muramic acid in soil. *Can. J. Microbiol.* **16**, 299–304.

MILLAR W.N. & CASIDA L.E. (1970b). Micro-organisms in soil as observed by staining with rhodamine-labelled lysozyme. *Can. J. Microbiol.* **16**, 305–307.

MILLER F.A. (1953). Applications of infrared and ultraviolet spectra to organic chemistry. In *Organic chemistry, an advanced treatise, III,* ed. H.Gilman. pp. 122–177. Wiley, New York.

MINDERMAN G. (1956). The preparation of microtome sections of unaltered soil for the study of soil organisms *in situ. Pl. Soil* **8**, 42–48.

MISHUSTIN E.N. (1955). Microbial associations of soils and methods of studying them. *Mikrobiologiya* **24**, 474–485 (in Russian).

NAGEL-DE-BOOIS H.M. (1971). Preliminary estimate of production of fungal mycelium in forest soil layers. *Rev. Ecol. Biol. Sol* **8** (in press).

NEWBOULD P.J. (1967). *Methods for estimating the primary production of forests.* IBP Handbook 2. Blackwell, Oxford.

NEWMAN A.S. & NORMAN A.G. (1943). An examination of thermal methods for following microbiological activity in soil. *Proc. Soil. Sci. Soc. Am.* **8**, 250–253.

NICHOLAS D.P. & PARKINSON D. (1967). A comparison of methods for assessing the amount of fungal mycelium in soil samples. *Pedobiologia* **7**, 23–41.

NIKITIN D.I. (1964). Use of electron microscopy in the study of soil suspensions and cultures of micro-organisms. *Soviet Soil Sci.* 636–641.

OKAFOR N. (1966). The ecology of micro-organisms on, and the decomposition of, insect wings in soil. *Pl. Soil* **25**, 211–237.

PARKINSON D. (1957). New methods for the qualitative and quantitative study of fungi in the rhizosphere. *Pedelogie, Gand.* **7** (no. spec.), 146–154.

PARKINSON D. & COUPS E. (1963). Microbial activity in a podzol. In *Soil organisms,* ed. J.Doeksen & J.van der Drift. pp. 167–175. North Holland Publishing Co., Amsterdam.

PARKINSON D., GRAY T.R.G., HOLDING J. & NAGEL-DE-BOOIS H.M. (1971). Heterotrophic microbiota. In IBP Handbook 18, ed. J.Phillipson, Blackwell, Oxford (in press).

PARKINSON D. & WILLIAMS S.T. (1961). A method for isolating fungi from soil microhabitats. *Pl. Soil* **13**, 347–355.

PATON A.M. (1959). An improved method for preparing pectate gels. *Nature, Lond.* **183**, 1812–1813.

PERFILIEV B.V. & GABE D.R. (1969). *Capillary methods of studying micro-organisms.* Oliver & Boyd, Edinburgh.

PETERSON R.G. & CALVIN L.D. (1965). Sampling. In *Methods of soil analysis. I,* ed. C.A. Black *et al.* pp. 54–72. American Society of Agronomy, Madison.

PIPER C.S. (1944). *Soil and plant analysis.* University of Adelaide Press, Adelaide.

PITAL A., JANOWITZ S.L., HUDAK C.E. & LEWIS E.E. (1966). Direct fluorescent labelling of micro-organisms as a possible life detection technique. *Appl. Microbiol.* **14**, 119–123.

POCHON J. (1957). Principe et application d'une methodologie quantiative. *Pedologie, Gand.* **7**, (no. spec.), 56–61.

PORTER L.K. (1965). Enzymes. In *Methods of soil analysis. II,* ed. C.A.Black *et al.* pp. 1536–1549. American Society of Agronomy, Madison.

REINERS W.A. (1968). Carbon dioxide evolution from the floor of three Minnesota forests. *Ecology* **49**, 471–483.

ROSSI G.M., RICCARDO S., GESUÉ G., STANGANELLA M. & WANG T.K. (1936). Direct microscopic and bacteriological examination of the soil. *Soil Sci.* **41**, 53–66.

RUBENCHIK L.I. (1963). *Azotobacter and its uses in agriculture.* Israel Program for Scientific Translations, Jerusalem.

SACKS L.E. & ALDERTON G. (1961). Behaviour of bacterial spores in aqueous polymer two-phase systems. *J. Bact.* **82**, 331–341.

SCHMIDT E.L. & BANKOLE R.O. (1963). The use of fluorescent antibody with the buried slide technique. In *Soil organisms,* ed. J.Doeksen & J.van der Drift. pp. 197–203. North Holland Publishing Co., Amsterdam.

SCOTT W.J. (1956). Water relations of food spoilage micro-organisms. *Adv. Fd. Res.* **7**, 83–127.

SKINNER F.A. (1951). A method for distinguishing between viable spores and mycelial fragments of actinomycetes in soil. *J. gen. Microbiol.* **5**, 159–166.

SKINNER F.A., JONES P.C.T. & MOLLISON J.E. (1952). A comparison of a direct- and a plate-counting technique for quantitative estimation of soil micro-organisms. *J. gen. Microbiol.* **6**, 261–271.

SMITH G. (1969). *An introduction to industrial mycology.* 6th ed. Arnold, London.

SMITH N.R. & WORDERN S. (1925). Plate counts of soil micro-organisms. *J. agric. Res.* **31**, 501–517.

SNEDECOR G.W. (1956). *Statistical methods.* Iowa State College Press, Ames, Iowa.

SPARROW F.K. (1957). *Aquatic phycomycetes.* University of Michigan Press, Ann Arbor.

STOTZKY G. (1965). Replica plating techniques for studying microbial interactions in soil. *Can. J. Microbiol.* **11**, 629–636.

STOTZKY G. (1965b). Microbial respiration. In *Methods of soil analysis, II*, ed. C.A.Black et al. pp. 1550–1570. American Society of Agronomy, Madison.

STOTZKY G., GOOS R.D. & TIMONIN M.I. (1962). Microbial changes occurring in soil as a result of storage. *Pl. Soil* **16**, 1–18.

STOVER R.H. & WAITE B.H. (1953). An improved method of isolating *Fusarium* spp. from plant tissue. *Phytopathology* **43**, 700–701.

TAYLOR H. (1967). *The use of glass microbeads in the development of an artificial soil-like system.* Ph.D. thesis, University of Waterloo.

TAYLOR J. (1962). The estimation of numbers of bacteria by tenfold dilution series. *J. appl. Bact.* **25**, 54–61.

TAYLOR L.R. (1961). Aggregation, variance and the mean. *Nature, Lond.* **189**, 732–735.

THORNTON H.G. (1922). On the development of a standardized agar medium for counting soil bacteria with especial regard to the repression of spreading colonies. *Ann. appl. Biol.* **9**, 241–274.

THORNTON R.H. (1952). The screened immersion plate. A method for isolating soil micro-organisms. *Research, Lond.* **5**, 190–191.

THORNTON R.H. (1958). A soil fungus trap. *Nature, Lond.* **182**, 1690.

TIMONIN M.I. (1941). The interaction of higher plants and soil micro-organisms. III. Effect of by-products of plant growth on activity of fungi and actinomycetes. *Soil Sci.* **52**, 395–413.

TRIBE H.T. (1957). Ecology of micro-organisms in soils as observed during their development upon buried cellulose film. In *Microbial ecology*, ed. C.C.Spicer & R.E.O.Williams. *Symp. Soc. gen. Microbiol.* **7**, 287–298. Cambridge University Press, Cambridge.

TRIBE H.T. (1963). The microbial component of humus. In *Soil organisms*, ed. J.Doeksen & J.van der Drift. pp. 205–211. North Holland Publishing Co., Amsterdam.

TYLDESLEY J.B. (1967). Movement of particles in the lower atmosphere. In *Airborne microbes*, ed. P.H.Gregory & J.L.Monteith. *Symp. Soc. gen. Microbiol.* **17**, 18–30. Cambridge University Press, Cambridge.

UMBREIT W.W., BURRIS R.H. & STAUFFER J.F. (1964). *Manometric techniques.* 4th edn. Burgess, Minneapolis.

VAN NIEL C.B. (1931). On the morphology and physiology of the purple and green sulphur bacteria. *Arch. Mikrobiol.* **3**, 1–112.

VAN NIEL C.B. (1944). Culture, general physiology, morphology and classification of the non-sulphur, purple and brown bacteria. *Bact. Rev.* **8**, 1–118.

WAID J.S. (1957). Distribution of fungi within the decomposing tissues of rye-grass roots. *Trans. br. mycol. Soc.* **40**, 391–406.

WALLIS G.W. & WILDE S.A. (1957). Rapid determination of carbon dioxide evolved from forests soils. *Ecology* **38**, 359–361.

WARCUP J.H. (1950). The soil plate method for isolation of fungi from soil. *Nature, Lond.* **166**, 117–118.

WARCUP J.H. (1951). The ecology of soil fungi. *Trans. br. mycol. Soc.* **34**, 376–399.

WARCUP J.H. (1955). Isolation of fungi from hyphae present in soil. *Nature, Lond.* **175**, 953–954.

WARCUP J.H. (1959). Studies on basidiomycetes in soil. *Trans. br. mycol. Soc.* **42**, 45–52.

WARCUP J.H. (1960). Methods for isolation and estimation of activity of fungi in soil. In *The ecology of soil fungi*, ed. D.Parkinson & J.S.Waid. pp. 3–21. Liverpool University Press, Liverpool.

WARCUP J.H. & BAKER K.F. (1963). Occurrence of dormant ascospores in soil. *Nature, Lond.* **197**, 1317–1318.

WASTIE R.L. (1961). Factors affecting competitive saprophytic colonization of the agar plate by various root-infecting fungi. *Trans. br. mycol. Soc.* **44**, 145–159.

WEBLEY D.M. & DUFF R.B. (1962). A technique for investigating localized microbial development in soils. *Nature, Lond.* **194**, 364–365.

WILLIAMS S.T., PARKINSON D. & BURGES N.A. (1965). An examination of the soil washing technique by its application to several soils. *Pl. Soil* **22**, 167–186.

WILLOUGHBY L.G. (1968). Aquatic Actinomycetales with particular reference to the Actinoplanaceae. *Ver. Inst. Meerforschung, Bremerhaven* **3**, 19–26.

WITKAMP M. (1966). Decomposition of leaf litter in relation to environment, microflora and microbial respiration. *Ecology* **47**, 194–201.

WITKAMP M. & CROSSLEY D.J. (1966). The role of arthropods and microflora in the breakdown of white oak litter. *Pedobiologia* **6**, 293–303.

WITKAMP M. & OLSON J.S. (1963). Breakdown of confined and non-confined litter. *Oikos* **14**, 138.

9283